Christian Matthew Winterflood

Applications of Supercritical Angle Fluorescence

Christian Matthew Winterflood

Applications of Supercritical Angle Fluorescence

Supercritical Angle Fluorescence-Based Microscopy, Correlation Spectroscopy, and Biosensing

Südwestdeutscher Verlag für Hochschulschriften

Impressum / Imprint
Bibliografische Information der Deutschen Nationalbibliothek: Die Deutsche Nationalbibliothek verzeichnet diese Publikation in der Deutschen Nationalbibliografie; detaillierte bibliografische Daten sind im Internet über http://dnb.d-nb.de abrufbar.
Alle in diesem Buch genannten Marken und Produktnamen unterliegen warenzeichen-, marken- oder patentrechtlichem Schutz bzw. sind Warenzeichen oder eingetragene Warenzeichen der jeweiligen Inhaber. Die Wiedergabe von Marken, Produktnamen, Gebrauchsnamen, Handelsnamen, Warenbezeichnungen u.s.w. in diesem Werk berechtigt auch ohne besondere Kennzeichnung nicht zu der Annahme, dass solche Namen im Sinne der Warenzeichen- und Markenschutzgesetzgebung als frei zu betrachten wären und daher von jedermann benutzt werden dürften.

Bibliographic information published by the Deutsche Nationalbibliothek: The Deutsche Nationalbibliothek lists this publication in the Deutsche Nationalbibliografie; detailed bibliographic data are available in the Internet at http://dnb.d-nb.de.
Any brand names and product names mentioned in this book are subject to trademark, brand or patent protection and are trademarks or registered trademarks of their respective holders. The use of brand names, product names, common names, trade names, product descriptions etc. even without a particular marking in this works is in no way to be construed to mean that such names may be regarded as unrestricted in respect of trademark and brand protection legislation and could thus be used by anyone.

Coverbild / Cover image: www.ingimage.com

Verlag / Publisher:
Südwestdeutscher Verlag für Hochschulschriften
ist ein Imprint der / is a trademark of
AV Akademikerverlag GmbH & Co. KG
Heinrich-Böcking-Str. 6-8, 66121 Saarbrücken, Deutschland / Germany
Email: info@svh-verlag.de

Herstellung: siehe letzte Seite /
Printed at: see last page
ISBN: 978-3-8381-3657-8

Zugl. / Approved by: Zürich, Universität Zürich, Diss., 2010

Copyright © 2013 AV Akademikerverlag GmbH & Co. KG
Alle Rechte vorbehalten. / All rights reserved. Saarbrücken 2013

*No pleasure is comparable to the standing upon the vantage ground of truth
…and to see the errors…in the vale below; always that this prospect be with pity,
and not with swelling or pride.*

In *The Essays, of Truth*, Sir Francis Bacon (1561–1626)

The research carried out in the context of this dissertation has resulted in contributions to the following publications and conferences:

Peer Reviewed Publications

- Christian M. Winterflood, Thomas Ruckstuhl, Dorinel Verdes, and Stefan Seeger.
  Nanometer Axial Resolution by Three-Dimensional Supercritical Angle Fluorescence Microscopy.
  Physical Review Letters, **105**(10), 108103 (2010).

- Thomas Ruckstuhl, Christian M. Winterflood, and Stefan Seeger.
  Supercritical Angle Fluorescence Immunoassay Platform.
  Analytical Chemistry, **83**, 2345 (2011).

- Thomas Ruckstuhl, Dorinel Verdes, Christian M. Winterflood, and Stefan Seeger.
  Simultaneous Near-field and Far-field Fluorescence Microscopy of Single Molecules.
  Optics Express, **17**(7), 6836 (2011).

- Christian M. Winterflood, Thomas Ruckstuhl, Nicholas P. Reynolds, and Stefan Seeger.
  Tackling Sample-Related Artifacts in Membrane FCS using Parallel SAF and UAF Detection.
  ChemPhysChem, **12**(16), 3655 (2012).

- Christian M. Winterflood, Thomas Ruckstuhl, and Stefan Seeger
  Fast and Sensitive Interferon-γ Assay Using Supercritical Angle Fluorescence.
  Biosensors, **3**(1), 108 (2013).

Presentations at Conferences

- Christian M. Winterflood, Thomas Ruckstuhl, Dorinel Verdes, and Stefan Seeger.
  Parallel Near- and Far-field Spectroscopy of Single Molecules.
  16th Picoquant International Workshop on Single Molecule Spectroscopy and Ultrasensitive Analysis in the Life Sciences, Berlin, Germany, September 15, 2010.

Posters at Conferences

- Christian M. Winterflood, Thomas Ruckstuhl, Dorinel Verdes, and Stefan Seeger.
  Three-Dimensional Supercritical Angle Fluorescence Microscopy.
  15th *Picoquant International Workshop on Single Molecule Spectroscopy and Ultrasensitive Analysis in the Life Sciences*, Berlin, Germany, September 7-9, 2009.

SUMMARY

Many applications of fluorescence in microscopy imaging, spectroscopy and biosensing rely on the sensitive and selective detection of fluorescence at an interface. Collecting fluorescence emitted above the critical angle, so-called supercritical angle fluorescence (SAF), provides a means for a highly surface-confined detection volume. At the same time a very high collection efficiency is achieved as a consequence of the high angles of fluorescence collection.

The present dissertation is divided into two parts covering developments in SAF microscopy and SAF-based biosensing.

The first part covers parallel near- and far-field microscopy. A novel method for sub-diffraction-limited resolution along the optical axis is established. It relies on the separate and simultaneous measurement of SAF and fluorescence emitted below the critical angle (undercritical angle fluorescence, UAF). UAF is weakly influenced by the emitter-surface distance and provides a measure for the intrinsic brightness of an emitter. Using a custom optical configuration, the parallel measurement of SAF and UAF has allowed for determining axial positions with a precision of more than two orders of magnitude below the diffraction-limit – even for single molecules.

The part will further deal with the development of fluorescence correlation spectroscopy (FCS) using both SAF and UAF simultaneously. Measurements for free diffusion are compared with simulations for characterizing both detection volumes. Additionally, diffusion experiments both in synthetic membranes and cell membranes are performed. A new FCS scheme is introduced to reduce sample-related artefacts in the determination of diffusion coefficients in cell membranes.

A section is devoted to the development of a second generation SAF microscope system. At its core is an objective with additional single molecule sensitivity of the collection optics for UAF. With this, simultaneous detection of both near- and far-field emission modes from single molecules without the use of a probe tip is demonstrated for the first time. The ability to detect UAF and SAF for single molecules at the same time opens up new venues for measuring fast dynamic processes at interfaces. The advanced microscope system allowed to extend the applications in microscopy imaging and FCS down to the single-molecule level.

The second part comprises the development and realization of rapid, robust, sensitive and costeffective immunoassays based on SAF using an in-house developed platform. The SAF technology allows for real-time monitoring of binding kinetics in a sandwich assay format. The assay is performed with single-use polymer test tubes and a compact reader and requires a minimal amount of material and working steps. Assays with picomolar sensitivity in less than fifteen minutes are presented for three popular analytes.

ZUSAMMENFASSUNG

Viele Anwendungen der Fluoreszenz in der mikroskopischen Bildgebung, Spektroskopie und Biosensorik setzen die sensitive und selektive Detektion von Fluoreszenz an Grenzflächen voraus. Das Sammeln der Fluoreszenz, welche über dem kritischen Winkel emittiert wird, sogenannte supercritical angle fluorescence (SAF), liefert eine Möglichkeit ein sehr nahe an einer Oberfläche lokalisiertes Detektionsvolumen zu erzielen. Gleichzeitig wird als Folge der hohen Sammungswinkel eine sehr grosse Sammlungseffizienz erreicht.

Die folgende Dissertation ist in zwei Teile unterteilt, welche Entwicklungen in der SAF Mikroskopie und der SAF-basierten Biosensorik umfassen.

Der erste Teil behandelt die parallele Nah- und Fernfeld Mikrosokopie. Eine neue Methode für eine Auflösung unter der Beugungsgrenze entlang der optischen Achse wurde etabliert. Sie basiert auf der getrennten und gleichzeitigen Messung von SAF und Fluoreszenz, welche unter dem kritischen Winkel emittiert wird (undercritical angle fluorescence, UAF). UAF ist nur schwach von der Entfernung Emitter-Oberfläche abhängig und dient als Mass für die intrinsische Helligkeit des Strahlers. Zusammen mit SAF konnten axiale Positionen mit einer Genauigkeit von über zwei Grössenordnungen unter der Beugungsgrenze bestimmt werden – sogar für Einzelmoleküle.

Weiter behandelt wird die Entwicklung von Fluoreszenzkorrelationsspektroskopie (FCS) mit SAF und UAF gleichzeitig. Für die Charakterisierung der beiden Detektionsvolumina werden FCS Messungen für freie Diffusion mit Simulationen verglichen. Zusätzlich werden Messungen der Diffusion sowohl in synthetischen Membranen, als auch in Zellmembranen gezeigt. Es wird eine neue Methode vorgestellt, um probenspezifische Artefakte in der Messung der Membrandiffusion in Zellen zu minimieren.

Weiter ist ein Teil der Entwicklung eines SAF Mikroskopaufbaus der zweiten Generation gewidmet. Herzstück des Mikroskops ist ein Objektiv mit zusätzlicher Einzelmolekülempfindlichkeit der Sammlungsoptik für UAF. Damit wird die gleichzeitige Detektion von Nah- und Fernfeld Emissionsmodi von einzelnen Molekülen für das erste Mal ohne Verwendung einer Probenspitze gezeigt. Die Fähigkeit SAF und UAF auf Einzelmolekülebene zu detektieren, eröffnet neue Möglichkeiten, um schnelle, dynamische Prozesse and Grenzflächen zu messen. Das weiterentwickelte Mikroskop Sytem ermöglichte die Anwendungen in der bildgebenden Mikroskopie und FCS auf Einzelmolekülebene auszuweiten.

Der zweite Teil umfasst die Entwicklung und Umsetzung schneller, zuverlässiger, empfindlicher und kosteneffektiver Immunoassays basierend auf SAF mittels einer intern entwickelten Plattform. Die SAF-Technolgie ermöglicht die Beobachtung von Bindungskinetiken in einem Sandwich-Assay-Format in Echtzeit. Der Assay wird mit Einweg-Teströhrchen und einem kompakten Gerät durchgeführt und erfordert ein Minimum an Material und Arbeitschritten. Assays mit pikomolarer Empfindlichkeit in weniger als fünfzehn Minuten werden für drei weitverbreitete Analyten vorgestellt.

Contents

1	**Introduction**	**1**
2	**Fluorescence Detection at Interfaces**	**5**
3	**Immunoassay Technologies**	**15**
4	**Optical Setups**	**21**
	4.1 Parallel Near- and Far-field Fluorescence Microscopes	21
	4.1.1 Prototype microscope	21
	4.1.2 Second generation microscope	22
	4.2 SAF Immunoassay Platform	34
5	**Parallel Near- and Far-Field Microscopy**	**39**
	5.1 Three-Dimensional Supercritical Angle Fluorescence Microscopy	40
	5.1.1 3D-SAFM measurements with the prototype microscope	43
	5.1.2 Axial localization with the *2-Theta* microscope	45
	5.2 Parallel Near- and Far-Field FCS	54
	5.2.1 Principles of FCS	54
	5.2.2 FCS simulations and measurements of diffusion in solution	62
	5.2.3 FCS measurements of diffusion in membranes	65
	5.3 Conclusions and Outlook	78
6	**Supercritical Angle Fluorescence Immunoassay Platform**	**81**
	6.1 Tube surface chemistry	81
	6.2 Assays	84
	6.3 Enhancement of the Assay Sensitivity	90
	6.4 Conclusions and Outlook	95
7	**Material and Methods**	**97**

8 Summary	105
Appendix	107
References	109
List of Abbreviations	119
List of Figures	123
Acknowledgements	127
Curriculum Vitae	129

1 Introduction

Many applications of fluorescence in microscopy imaging, biology, medical research and diagnosis require the sensitive and selective detection of fluorescence near interfaces. The fluorescence detection from only a thin focal plane is essential for specimens where fluorescence information can be obscured by the presence of a large number of fluorophores located outside of the optical plane of interest. This is for instance the case when studying cell-substrate contacts or molecular interactions at surfaces. Optical near-fields can be used to spatially confine the fluorescence detection to volumes of sub-wavelength extension at an interface. Here, a near-field approach is adopted, which makes use of the directionality of fluorescence emission at an optical boundary, e.g. a water/coverslip-glass interface. The perturbation of a fluorescing emitter's near-field created by the proximity of the refractive index (RI) jump leads the creation of detectable propagating waves in the far-field and a significant part of the radiation is emitted at angles above the critical angle. SAF accounts for more than one third of the overall emission and is exponentially dependent on the distance from the glass. A very surface-confined detection volume together with a high collection efficiency can be achieved by the exclusive collection of SAF. The surface selectivity is achieved by supercritical fluorescence collection rather than excitation as in total internal reflection fluorescene (TIRF) microscopy. The surface can therefore be illuminated at angles below the critical angle for the exciting beam to propagate into the specimen. This offers the possibility to detect fluorescence in deeper axial planes by additionally collecting the far-field mediated emission below the critical angle, termed UAF. A customized optical geometry is used for the well defined separation of UAF and SAF. The geometry comprises an inner optics used for point illumination and for the collection of UAF surrounded by a parabolic collector for the collection of SAF. Parabolic collectors are highly efficient elements for the collection of fluorescence at high surface angles. With this optical geometry parallel detection of near- and far-field was shown the first time without using a probe tip.[1] Hecht *et al.* have demonstrated it in the context of near-field scanning optical microscopy.[2]

The parallel measurement of SAF and UAF will be the basis of a newly introduced fluorescence microscopy technique for sub-diffraction-limited resolution along the optical axis. The limit for optical techniques set by diffraction (200 nm in the lateral and 500 nm in the axial direction for

1

a standard confocal fluorescence microscope) is larger than many subcellular structures, leaving them too small to be observed in detail. This gap has been the driving force in pursuit of so-called super-resolution – a resolution which is not limited by diffraction. In recent years, a number of sub-diffraction fluorescence microscopy techniques have emerged which are beginning to revolutionize our understanding in cellular biology. Of these, stimulated emission depletion microscopy (STED) and single-molecule localization based methods, such as stochastic optical reconstruction microscopy (STORM) and photoactivated localization microscopy (PALM), have become the most established techniques but *per se* provide sub-diffraction resolution only in the lateral dimension. In the quest of 3D nanoscale resolution, the axial dimension needs to be specifically addressed because of the symmetry of the implementation setups. Several techniques have been combined with STED and STORM/PALM for 3D sub-diffraction resolution – some technically very demanding[3–5] or with fundamental limitations.[6,7] Exceptionally high axial resolutions have been achieved by using the properties of optical near-fields and their ability to localize optical energy to length scales smaller than the diffraction-limit. By scanning the specimen with a light source or aperture of sub-wavelength extension as in NSOM, resolutions of less than 20 nm can be achieved in all three dimensions.[8] Due to the requirement of a tip NSOM is, however, limited to the study of surfaces and non-invasive near-field approaches have been proposed. Some techniques make use of TIRF illumination to measure axial positions with less than 10 nm accuracy.[9,10] These methods rely on precise calibration of the evanescent field depth and even more on stable fluorescence emitters hampering their use for z-localization of fluctuating probes, such as single fluorophores, photoswitchable labels, and quantum dots. In a method developed in this work the simultaneous detection of SAF and UAF makes it possible to account for fluorescence intensity fluctuations of small emitters on any time scale. Axial positions are determined non-invasively with a precision of more than two orders of magnitude below the diffraction-limit. The technique is in principle compatible with STED and STORM/PALM techniques.

The applications and opportunities offered by parallel detection of near- and far-field are extended with the development of a second generation microscope system in this work. The advanced setup now provides single-molecule sensitivity also for UAF emission modes allowing for sub-diffraction localization of single molecules by detecting their near- and far-field.

The innovative optical configuration employed has some powerful benefits over other techniques for surface-confined FCS. FCS is a highly established technique used to measure a number of physical quantities including concentrations, diffusion coeffcients, binding equilibria and binding kinetics. One of the challenges in FCS is the restriction of the observation volume to reject stray light, autofluorescence and for the use of high concentrations typical for *in vivo* processes. For the study of processes at surfaces, the standard confocal FCS has the immanent problem that the ellipsoidal detection volume suffers from surface selectivity having a height-to-diameter ratio of 3–5. Surface processes remain concealed by the background produced by the bulk fluorescence. Surface-confined FCS has been achieved with evanescent waves produced at optical nanostructures called zero-mode waveguides[11] or more commonly using TIRF.[12–14] In contrast to SAF, the penetration depth of the created evanescent wave with TIRF is highly dependent on the illumination angle. Precise laser beam angles are usually not known *a priori* and quantitative FCS results are often very difficult to obtain. In TIR-FCS for the measurement of slow diffusion at interfaces, e.g. membrane diffusion, out-of-focus photobleaching poses a serious problem. Many fluorophores get bleached in the large illuminated surface before they transit the detection area defined by a pinhole.

The SAF approach has shown to be well-suited for the study of surface-near processes and membrane diffusion.[15] The presented optical configuration allows for precise control of collection angles for SAF, providing a well defined detection volume along the optical axis. The illumination at moderate surface-angles affords a small excitation spot of Gaussian shape which minimizes the effect of out-of-focus photobleaching and simplifies the mathematical interpretation of FCS curves.

FCS is implemented on the newly developed microscope which now affords the necessary sensitivity to perform standard confocal FCS using UAF (UA-FCS) and surface-confined FCS using SAF (SA-FCS) for the first time simultaneously. This is a powerful combination for the study of dynamic processes occurring at interfaces and membranes. A thorough characterization of the SAF and UAF detection volumes by FCS experiments is provided for the new microscope system. A comparison with simulations and theoretical models is made. Diffusion measurements are performed on artifical membranes and the plasma membrane of cells and a method is introduced for minimizing sample-related artefacts in the determination of diffusion coefficients by FCS in cell membranes. The method relies on the axial sensitivity provided by the parallel detection of near- and far-field emission modes.

The SAF technique has shown to be particularly useful for following binding reactions at interfaces.[16–26] Especially the simple parabolic geometry combined with the extraordinarily high collection efficiency for surface-bound fluorescence provides the ideal foundation for an immunoassay platform for the quantification of analytes. Highly sensitive techniques rely on a heterogeneous assay format, where analyte molecules accumulate on a solid substrate. Label-free approaches are based on the direct measurement of a phenomena occurring during the immunoreaction on a transducer surface. These techniques are based on sensing gravimetric changes using a quartz-crystal microbalance[27] or a microcantilever,[28] changes in the electrical conductance of nanowires and nanotubes,[29] thermal changes as in nanocalorimetry,[30] and surface plasmon resonance.[31,32]

Labelled assay types, on the other hand, rely on the detection of a signal produced by the label of a query molecule. Labelled methods include radioimmunoassays, enzyme-linked immunosorbent assays, protein microarrays and quantum dot detection platforms, where the readout is based on a radioactive,[33] fluorescent[34,35] or colorimetric signal.[36] Other labelled types are based on the detection of surface enhanced Raman scattering[37] and electro-chemiluminescence.[38,39] The use of labels poses some synthetic challenges as well as the possible issue of interference of the tagging molecule. However, label–free detection techniques are generally less sensitive and display a lower selectivity. Many of the aforementioned techniques achieve high sensitivies, but for some the level of technical complexity and the associated costs restricts their use to centralized laboratories and research institutions. The well established enzyme-linked immunosorbent assay (ELISA) therefore still remains one of the most widely used detection platforms for the quantification of bioanalytes. On the detections side, the ELISA uses a multitude of expensive substances. In a typical configuration, it requires a biotinylated detection antibody, a streptavidin-enzyme conjugate, and an enzyme substrate. The assay procedure involves numerous washing steps and time-consuming incubation periods amounting in total to several hours if not even requiring an over night incubation. There is clearly a need for a simpler and more robust alternative to the time-, work- and cost-consuming ELISA.

In this consideration, a part of this work focuses on the development of a low-cost, fluorescence-based immunoassay platform for rapid and sensitive immunoassays with potential use in point-of-care diagnostics. The system works with single-use polymer test tubes with incorporated optics and a small fluorescence reader. The solid-phase immunoassay is based on the detection of SAF. The assay is performed in a sandwich assay format where fluorescently labelled detection antibodies accumulating at the transparent polymer interface upon formation of sandwich complexes emit SAF. A parabolic collector converts the SAF into conveniently detectable parallel rays. Binding-kinetics are monitored without interference of the bulk fluorescence in solution. Therefore, the technique requires for no washing steps. In addition, due to the high collection efficiency of the parabolic approach no biochemical signal amplification is necessary. Assays with picomolar sensitivity within 13 min are presented for three common analytes.

2 Fluorescence Detection at Interfaces

This chapter is concerned with providing an overview of surface fluorescence techniques. Although not usually referred to as surface techniques, confocal fluorescence and wide-field epi-fluorescence microscopy also deserve mention, being restricted to imaging only a few hundreds of microns above the microscope slide. The chapter is concluded with the discussion of two specific sub-diffraction microscopy techniques – due to their importance and inasmuch they can complement the techniques introduced in chapter 5.

Total internal reflection fluorescence

TIRF microscopy makes use of an optical near-field produced upon interaction of a transparent surface with free radiation from the far-field. The non-propagating evanescent field is produced upon illumination of the sample/glass interface above the critical angle of total internal reflection θ_c given by Snell's law as θ_c=arcsin(n_1/n_2) with ($n_2 > n_1$). Its intensity decays exponentially from the interface to define a axially sub-diffraction limited excitation volume [Fig. 2.1A]. The intensity of the evanescent field in z is given by[15]

$$I(z) = I_0 e^{-2w(\theta)z} \qquad (2.1)$$

with

$$w(\theta) = \frac{2\pi}{\lambda}\sqrt{n_2^2 \sin(\theta)^2 - n_1^2}, \qquad (2.2)$$

where I_0 is the intensity at z=0, and $w(\theta)$ is the incidence angle-dependent decay-length – more commonly denoted by d. The excitation volume of the TIRF evanescence field extends only ~100 nm (e.g. for λ=488 nm, θ=66°, n_1=1.333 (water), n_2=1.523 (glass)) into the sample to excite fluorophores only in that thin axial section adjacent to the sample/glass interface. Two common approaches to TIRF microscopy exist, often referred to as prism-type and objective-type, each illustrated in Fig. 2.1B. In prism-type TIRF, a focused laser beam is introduced into the microscope coverslip above θ_c using a prism attached to its surface. The prism technique is easy to set-up, but has

CHAPTER 2. FLUORESCENCE DETECTION AT INTERFACES

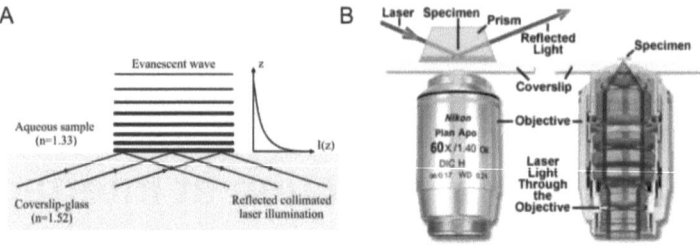

Figure 2.1: Principle of TIRF microscopy where (A) illumination above θ_c of the interface creates an exponentially decaying evanescent field with a decay length in the order of ∼100 nm ($1/e$ intensity). (B) Prism-type (*left*) and objective-type (*right*) specimen illumination configurations. Figure B adapted from Ref.[40]

the severe limitation of restricted access to the specimen. Making manipulations and changing the experimental conditions is very diffucult. In the objective-type TIRF, the laser is introduced through the objective to illuminate the coverslip/specimen interface at supercritical angles. This requires the use of objectives of high NA – typically 1.49. The objective-type TIRF bears the key advantage that the sample is accessible from above and has become the most popular method.

Supercritical angle fluorescence

The angular distribution of radiation of a fluorophore within around one emission wavelength from a planar dielectric interface is significantly perturbed.[41–43] An emitter located in the medium with lower RI emits a substantial proportion of its light into the medium with higher RI above θ_c [Fig. 2.2]. The distribution of radiation has a pronounced maximum in the direction of θ_c. Classically, no light incident from the side with the lower RI can enter angles above θ_c. Only fluorophores close enough for their near-field to couple with the interface emit this so-called forbidden light or SAF. As a consequence, a high surface selectivity of fluorescence collection can be achieved by exclusively collecting SAF. For a randomly oriented fluorophore at a water/glass interface SAF accounts for a formidable 34% of the radiated power and decays to less than 1% within only one wavelength. In contrast, the fluorescence emitted into undercritical angles (UAF) is only moderately affected by the distance from the surface [Fig. 2.3]. The SAF approach achieves a surface confinement comparable to TIRF but through collection rather than excitation above θ_c. A comparison of the penetration depths of the observation volumes is shown in Fig. 2.4. The TIRF method achieves a comparable resolution to SAF only with very large illumination angles, which are technically dif-

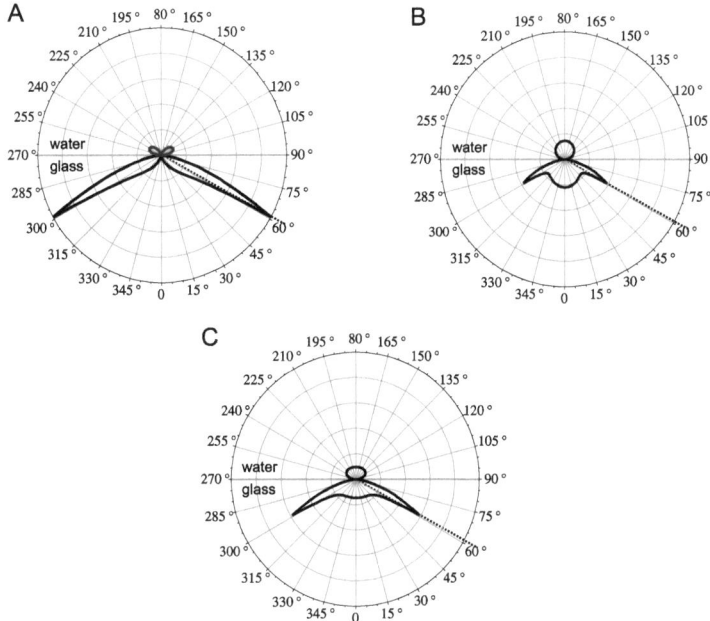

Figure 2.2: Polar plots of the radiant intensity for emitters with distance $z=0$ from a water/glass interface $n_1=1.333$, $n_1=1.523$) (A) with a dipole orientation perpendicular, (B) a dipole orientation parallel to the surface and (C) a randomized orientation. The *dashed lines* designate $\theta_c=61.1°$. SAF is the proportion of fluorescence emitted above θ_c into the glass. Note, that the emission distribution is not rotationally symmetric about the azimuthal angle for the parallel orientation. Here, it is shown for a dipole pointing out of the image plane.

ficult to realize. It is important to note, that each supercritical collection angle θ is associated with an exponential decay $e^{-2w(\theta)z}$ along the optical axis which is identical for the corresponding angle of TIRF excitation[15] (see Eqs. 2.1 and 2.2). Besides the angular emission distribution, the lifetime of an excited state, or expressed differently, the emission rate of a dipole is also affected by the presence of a dielectric interface. The overall radiated power of a fluorophore is thus altered in the case of excitation saturation. But because excitation saturation is generally avoided during experiments, the angular emission distributions shown in Fig. 2.3A and the relative SAF intensity shown in Fig. 2.3B were normalized by the total radiated power. The theory of fluorescence emission at planar dielectric interfaces is provided in the appendix on page 8.

CHAPTER 2. FLUORESCENCE DETECTION AT INTERFACES

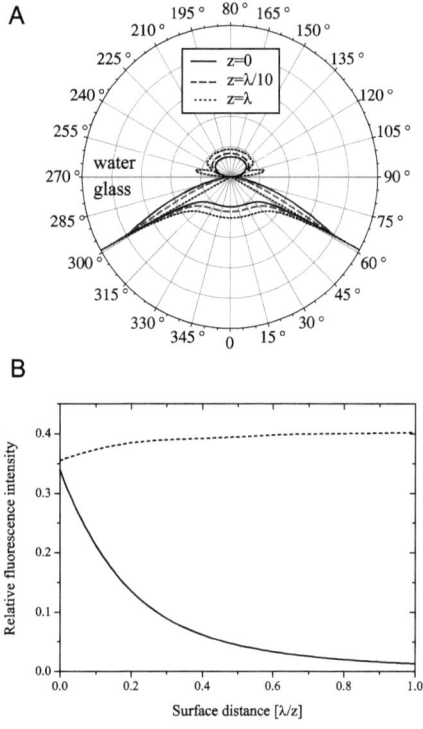

Figure 2.3: (A) Polar plots of the radiant intensity for emitters with distances $z=0$, $z=\lambda/10$, $z=\lambda$ and (B) the z-dependence of the relative SAF intensity (*red line*) and in comparison the relative UAF intensity emitted into the glass semi-space (*blue line*) calculated for a water/glass interface.

Near-field scanning optical microscopy

In NSOM high-resolution information is obtained by detecting non-propagating light near the fluorophore. A sub-wavelength light source is positioned in close proximity to the specimen. An image is aquired by scanning the probe tip across the sample (the tip is kept at constant height from the sample) providing resolutions far below the diffraction-limit in the order of 10-50 nm.[44] A representation of this most common imaging scheme – transmission mode NSOM – is presented in Fig. 2.5. While NSOM affords high spatial, spectral, and even chemical resolution in combination with laser-ablation and mass spectrometry,[45] it is inherently limited to the study of surfaces.

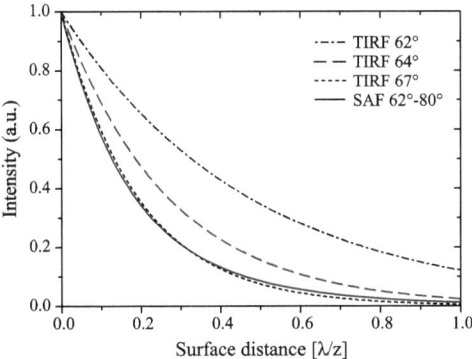

Figure 2.4: Comparison between the surface selectivity of SAF and TIRF. Instensity decays of the evanescent wave by TIRF illumination at 62°, 64°, and 67° and the collection efficiency of SAF for the typically collected angular range between 62° and 80 ° for water-glass. The TIRF decay approaches the SAF decay only for a comparably high excitation angle of 67°. The emission wavelength of SAF was multiplied by 1.05 to account for the Stokes shift.

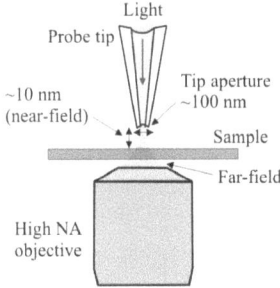

Figure 2.5: Principle of transmission mode NSOM.

Zero-mode wave guides

Zero-mode waveguides (ZMWs), although not used in imaging microscopy, serve to produce arrays of tiny sub-diffraction detection volumes. They consist of nanostructured, typically cylindrical wells with ∼50 nm diameter and ∼100 nm height in a thin aluminium film used for excitation volume confinement. As a result of their sub-wavelength size, they show no optical propagation modes. Illumination of the apertures results in an evanescent field within the aperture confining the excitation volume to atto- or even zeptoliters. Because of the small sampled volume, ZMWs offer the unique advantage of studying single molecule events at physiologically relevant molecular concentrations (>1 μM), such as required for DNA sequencing or FCS for measuring weak interactions.[46]

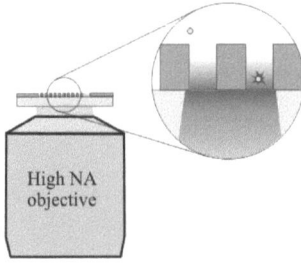

Figure 2.6: Zero mode waveguides.

Wide-field and confocal microscopy

In conventional wide-field epi-fluorescence microscopy the entire specimen is flooded evenly with light and all parts of the specimen in the optical are excited at once. The fluorescence emission is detected on a camera including a large part from unfocused background below and above the focal plane [2.7A]. By contrast, confocal microscopy uses a tightly focused laser and a pinhole in an optically conjugate plane in front of the detector to reject out-of-focus fluorescence [2.7B]. It is possible to exclusively image a thin optical slice out of a thicker specimen. As only a point in the sample is illuminated at the same time, the formation of an image requires scanning over a regular raster.

Super-resolution techniques

The resolving power of an optical system is fundamentally limited by the wave nature of light. In the imaging process of a lens based microscope, light rays from a point in object space converge to a point in the image plane. Instead of converging to a infinitely sharp spot, diffraction and interference cause the point on the object to spread in the image plane. The intensity distribution in all three dimensions of this spot is termed the point-spread function (PSF) which determines the resolving capacity of the microscope. Two points separated by less than the full width at half-maximum (FWHM) of the PSF overlap substantially in the image plane and are difficult to resolve.

CHAPTER 2. FLUORESCENCE DETECTION AT INTERFACES

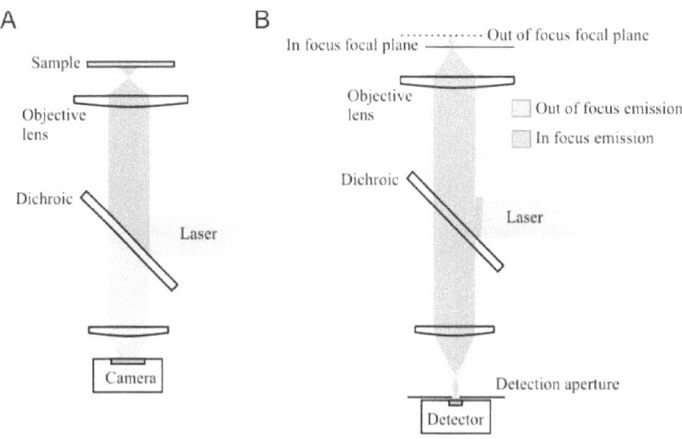

Figure 2.7: Far-field microscopy techniques. (A) Wide-field and (B) confocal fluorescence microscopy.

The FWHM of the PSF is given by $\Delta r \approx \lambda/(2n\sin(\alpha))$ in the focal plane and by $\Delta z \approx \lambda/(n\sin^2(\alpha))$ along the optical axis, where λ, α, and n denote the wavelength, the aperture half-angle of the objective, and the RI of the imaging medium, respectively.[47] The combined term $n\sin(\alpha)$) is known as the numerical aperture (NA). In practice, this means a resolution of ~200 nm in xy and ~500 nm in z when imaging with visible light ($\lambda \approx 550$ nm) and commonly used oil immersion objectives with NA=1.40.

For some time, only the use of the near-field provided an optical means for observing objects beyond this resolution. Only the recent two decades have shown a tremendous growth in optical techniques for sub-diffraction resolution microscopy which make use of the far-field. These non-invasive methods have opened up unprecedented new possibilities for investigating the structure and function of cells. The most important are covered in detail by several good review articles.[47-49] Stimulated emission depletion (STED) microscopy was the first to achieve super-resolution by far-field optics[50] and has been followed by ground-state depletion,[51] structured illumination microscopy (SIM),[52] and image interference microscopy.[53] More recently, stochastic techniques have been developed based on the analysis of temporal fluorescence fluctuations as in stochastic optical fluctuation imaging (SOFI)[54] or on single-molecule localization of photoswitchable probes, such as fluorescence photo-activated localization microscopy (fPALM),[55] stochastic optical reconstruction microscopy (STORM),[56] PALM[57] and variants thereof.[58,59] Other new concepts are based on

11

the modulation of the excitation intensity distribution as in image scanning microscopy (ISM)[60] or saturation of the excited state as in dynamic saturation optical microscopy (DSOM).[61] Of these methods, STED and STORM/PALM have the highest resolutions and have even become commercially available. But *per se* they provide sub-diffraction resolution only in the lateral dimension. STED microscopy relies on producing sub-diffraction limited features in the excited state population upon read-out of the fluorescence signal. Instead of returning from the excited state to the ground state

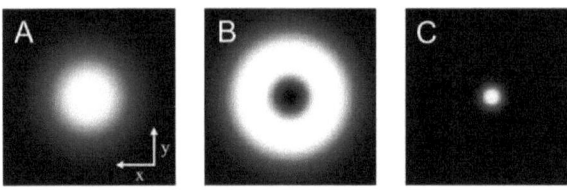

Figure 2.8: (A) Excitation spot, (B) doughnut-shaped depletion spot and (C) effective fluorescent spot.

by spontaneous emission, fluorophores can be quenched to the ground state upon encountering another photon with a wavelength comparable to the energy difference between the ground and excited state. In STED, this is used by superimposing the excitation beam with a doughnut-shaped depletion beam to quench excited molecules at the periphery of the excitation spot, thereby restricting emission to the doghnut-zero [Fig. 2.8]. Lateral resolutions down to 20 nm have been reported using STED.[62] Although the image of a single fluorophore is given by the PSF, the precision of determining the fluorophore position from the centroid of its image can be much higher. In STORM and PALM a high resolution image (<20 nm) is generated by consecutive localization of sparse subsets of fluorphores which are separated by at least the diffraction-limited resolution of the imaging instrument. Photo-activation is used to image only a fraction of the fluorophores in the field of view during every imaging cycle. The concept is illustrated in Fig. 2.9. STED and STORM/PALM have been combined with several other approaches for sub-diffraction-limited resolution additionally along the optical axis, but often either with a relatively poor axial resolution,[6,7] or at the expense of major technical demands or modifications to the microscope.[3-5]

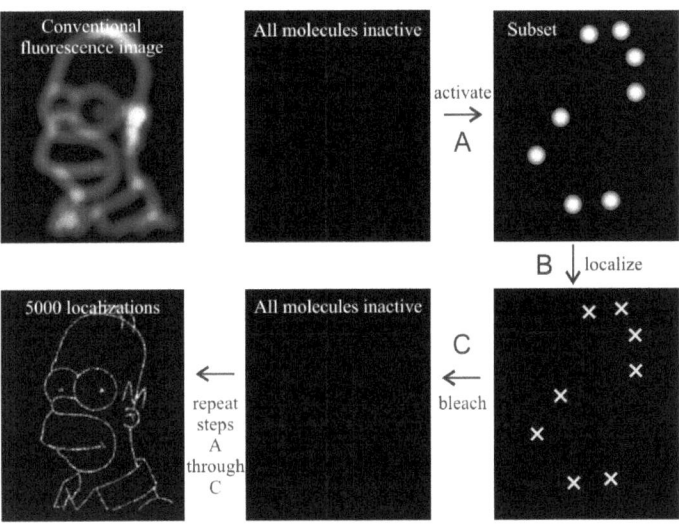

Figure 2.9: Demonstration of PALM on a richly structured sample (portrait of the cartoon character and American hero Homer Jay Simpson).

3 Immunoassay Technologies

This chapter will provide a brief overview of some of the state-of-the-art immunoassay technologies. Immunoassays have become a major scientific interest with a wide variety of biomedical, environmental and homeland security applications. Key factors determining the success of an immunoassay are sensitivity, dynamic range, real-time capability, the possibility of multiplexing, and general applicability. Common to all methods is that the analyte to quantify is known to undergo a unique immune reaction with a second species. This selective reaction is then used to produce a measurable signal for determining the presence and for quantifying the analyte in question. As there is a myriad of biochemical strategies for targeted analyte delivery, analyte separation and signal amplification, this part will only focus on some of the most important types of signal transduction.

Electrochemiluminescence

Electrochemiluminescence (ECL), as its name suggests, is based on chemiluminescence as a result of a series of electric current-driven energetic electron transfer (redox) reactions. In its most popular realization the ruthenium complex $Ru(bpy)_3$ (bpy = 2,2′-Bipyridine) is used as the light-emitting label molecule (e.g. the label for an antibody) and TPA (Tripropylamine) as the coreactant. The working principle is illustrated in Fig. 3.1. In the commonly used sandwich format a capture antibody is immobilized onto an anode. The $Ru(bpy)_3$ is brought to the vicinity of the anode upon formation of the sandwich complex with the antigen. The $Ru(bpy)_3$ and TPA are oxidized at the surface of the electrode upon application of a voltage. The TPA loses a proton reducing the ruthenium complex to an excited state. This relaxes to the ground state upon emission of a photon of ∼620 nm. With the $Ru(bpy)_3$ tag not beeing consumed, this cycle can be repeated indefinitely as long as TPA is present. Multiple excitation/emission cycles amplify the emitted light thereby increasing the sensitivity. The light intensity relates to the concentration of the analyte in the sample. A major advantage of this approach is an extremely low background as the detected optical signal is decoupled from the electrical stimulation. Using ECL based techniques, sub-picomolar

CHAPTER 3. IMMUNOASSAY TECHNOLOGIES

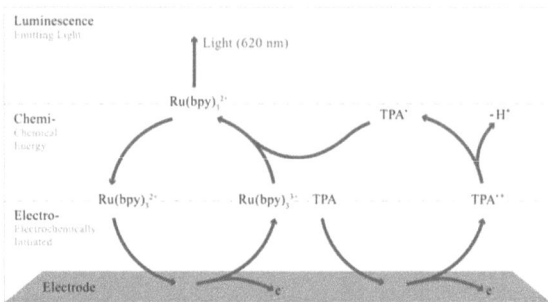

Figure 3.1: Basic principle of the ECL technology.

sensitivities and dynamic ranges of over three orders of magnitude have been reported.[63] ECL bears the disadvantage of costly instrumentation for automation and is generally restricted to the measurement of single analytes.

Total internal reflection fluorescence

In optical biosensing an important prerequisite is a small detection volume confined to the region where the immunoreaction takes place to reject contributions from background. TIRF is a very common approach to restrict fluorescence collection to only the sensor surface. As such, the binding reaction at the interface can be monitored in real-time without the influence of the excess unbound fluorescent detection species. Figure 3.2 shows an example of an implementation of TIRF for analyzing the binding of proteins to DNA.[64] A strong benefit of this approach is the ability

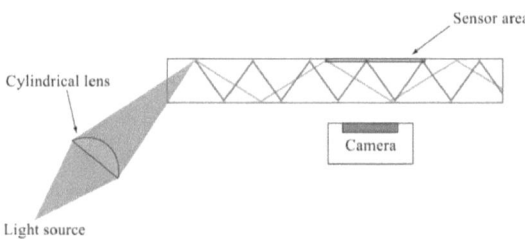

Figure 3.2: An evanescent excitation wave is generated using the microscope slide as an optical waveguide. Fluorescence in multiple excitation/emission pairs is imaged in real-time on a CCD-camera during the binding reaction, giving equilibrium and kinetic measurements

of multiplexing a large number of analytes. However, a high sensitivity requires expensive high aperture collection optics and sensitive CCD camera. An alternative TIR approach is the use of frustrated total internal reflection (f-TIR). If the evanescent wave produced by TIR extends across a separating medium into a region occupied by a higher RI material, the evanescent wave is frustrated and energy may flow across the boundary. A cost-effective method based on this principle is shown in Fig. 3.3. In this example sub-picomolar sensitivies with a measurement range of three orders of magnitued were achieved.[65]

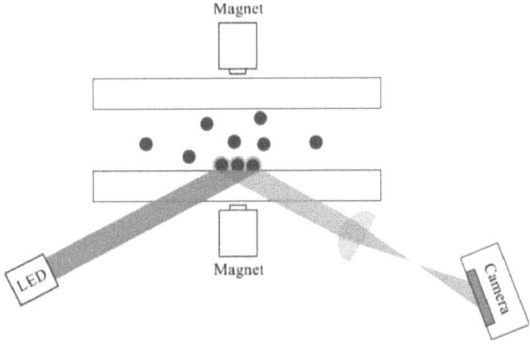

Figure 3.3: Antibody coated nanoparticles form sandwich complexes on the sensor surface with the antigen. Free and weakly bound nanoparticles are removed by magnetic actuation. The light reflected from the sensor surface depends inversely on the concentration of nanoparticles at the sensor surface.

Electrical biosensors

Nanowire field-effect transistors show a large potential for electronic detection of biomolecules in solution. In field-effect transistors the conductance is dependent on gate voltage. The selective binding of a charged antigen to the silicon nanowire gate dielectric results in an electric field which is analogous to applying a voltage using a gate electrode [Fig. 3.4]. This label-free approach allows for large scale parallelization and cost-effective sensor production by established semiconductor industry processes. However, in general electrical sensors based on carbon nanotubes or silicon nanowires do not demonstrate extremely high sensitivity (high picomolar range), but there are cases in which sub-picomolar detection limits have been reported.[29]

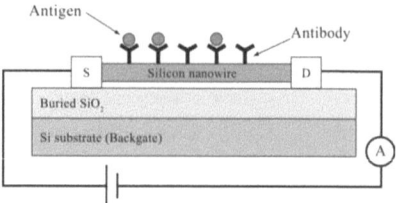

Figure 3.4: Schematic of a nanowire field-effect transistor immunoassay.

Surface plasmon resonance

Surface plasmon resonance (SPR) is a widely used method for analyzing receptor-ligand interactions. The principle of SPR is illustrated in Fig. 3.5. The measurement is based on the interaction of light with thin metal films on glass – typically a 40–50 nm thick gold layer. Light interacting with the gold layer has a very low reflectivity only around the SPR angle. This SPR angle is highly dependent on the RI, which can change upon accumulation of analyte at the receptor-coated interface. In SPR the increase in the reflectivity is measured as the SPR angle shifts. SPR is used in particular for determining binding rates being a label-free (no influence of the tag) and real-time method. It is becoming increasingly used as a means also for the quantification of analytes. However, lower pg/ml concentration require the use of nanoparticles for signal amplification.[66]

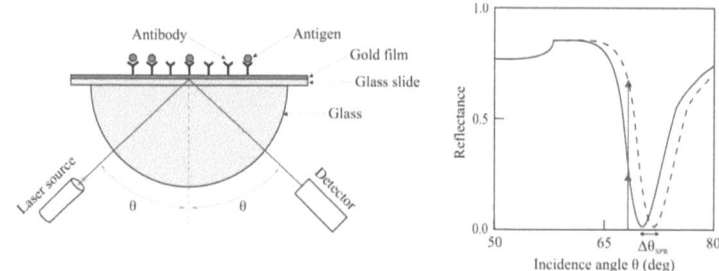

Figure 3.5: Principle of surface plasmon resonance.

Mass sensing

A mechanical approach to biosensing using microcantilevers has recently gained large interest. Microcantilevers are the most ubiquitous and well-studied structures in the field of microelectromechanical systems. A cantilever is a rectangular thin beam tethered at one end. Its upper surface is chemically modified so that it can react with specific compounds. The cantilever deflection, commonly measured by an optical beam deflection system, is a senstive measure of the mass loading from receptor antigen interaction.[67] Another approach for sensing minute changes in mass uses

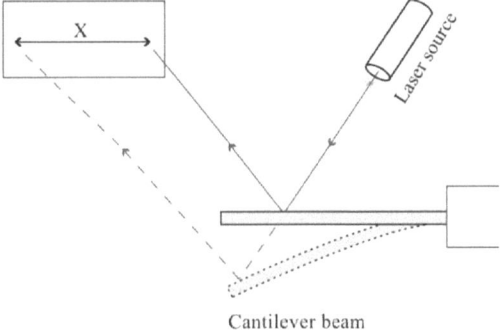

Figure 3.6: Microcantilever immunosensor.

quartz crystal microbalances (QCMs). QCMs are resonators with highly precise and stable natural resonant frequencies. The crucial component of the microbalance, the quartz crystal, registers and reports the mass deposited on its electrodes quantitatively. The mass load changes the frequency and the dampening of the oscillating piezoelectric crystal.[68]
With detection limits in the order of ~100 pg/ml these label-free mass sensing approaches are less sensitive, though offer the possibility of parallelization.

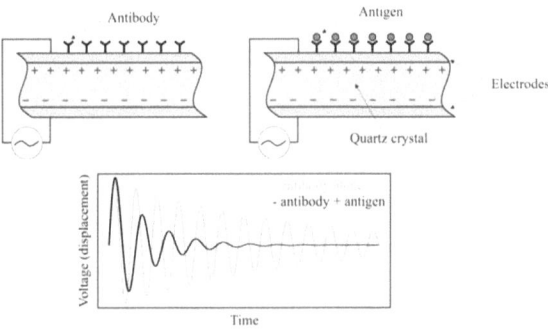

Figure 3.7: Quartz crystal microbalance immunosensor.

19

4 Optical Setups

During this work two different near- and far-field fluorescence microscope setups were used. A larger part of this chapter is devoted to the development and performance characteristics of the microscope of the second generation. The last part describes the SAF immunoassay platform.

4.1 Parallel Near- and Far-field Fluorescence Microscopes

With a NA of 0.6, the optical performance of the inner optics of the prototype SAF objective does not provide single molecule sensitivity for UAF. A part of this section will deal with the development of a second generation SAF objective and microscope setup with additional single-molecule sensitivity for UAF. The inner optics for UAF newly comprises a multilens system with a NA of 1.0 capable of diffraction limited performance at several different wavelengths. The first setup is reported capable of single-molecule sensitive detection of both near- and far-field emission modes in parallel using a single objective. The advanced system is used for the axial localization of single molecules and the axial tracking of diffusing nanobeads (see section 5.1.2), as well as for parallel FCS (see section 5.2).

4.1.1 Prototype microscope

A thorough description of the prototoype microscope shown in Fig. 4.1 can be found in Ref.[1,69] The main technical aspects and performance of the prototype will be briefly discussed here to provide a comparison with the second generation system. The inner optics of the objective consists of a single asphere with NA 0.6 (collected angles: 0 – 24°) embedded in a parabolic collector (collected angles: 62 – 75°). This corresponds to a collection efficiency of ∼6% for the aspheric and ∼30% for the parabolic collector for fluorescence at the water/coverslip interface. The parabolic surface of the element acts as a loss-free mirror via total internal reflection at the parabola/air interface. A pulsed 635 nm circular polarized diode laser is used as excitation source. The asphere is designed to focus the laser onto the surface of a # 1 coverslip, which is connected to the planar surface of the parabola with a submicron film of immersion-oil. The vertical position of the coverlsip is kept constant by

CHAPTER 4. OPTICAL SETUPS

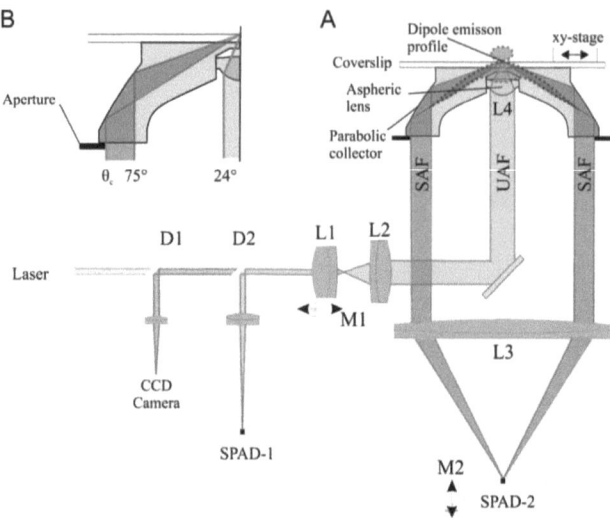

Figure 4.1: *(A)* Schematic of the prototype SAF microscope adapted from.[69] *L1 to L3* lenses, *M1 and M2* motorized tranlsational stages, and *SPAD-1(-2)* single photon avalanche diodes. *(B)* Enlarged view of the optical paths of the objective. Figure adapted from Ref.[1]

capillary for the scanning of sizeable areas. The laser focus is positioned to the coverslip surface by moving an expander lens with a motorized stage along the optical axis. A second motorized stage moves the SAF detector to the position of SAF image point along the optical axis depending on the coverslip thickness. The focus has a lateral and axial extent of 433 nm and 2.5 µm ($1\backslash e^2$-intensity drop-off), respectively.

4.1.2 Second generation microscope

The *2-Theta* objective

A new microscope objective is introduced for the implementation of parallel near- and far-field microscopy with single-molecule sensitivity on common microscope bodies. We call the objective *2-Theta* in reference to the two separate angular regions for fluorescence collection. The infinity corrected oil immersion microscope objective shown in Fig. 4.3 is designed for # 1.5 coverslips. It comprises an inner, custom-built multilens NA 1.0 system corrected for the prominent wavelengths 514 nm, 532 nm and 635 nm and a surrounding parabolic collector. The inner optics collects UAF

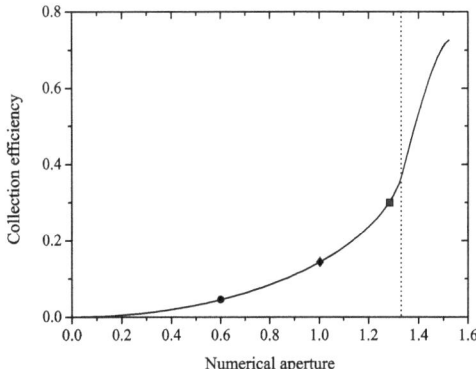

Figure 4.2: CE of a microscope objective in dependence of the NA with the CE of the prototype inner optics (*circle*) and the *2-Theta* inner optics (*diamond*). The CEs were calculated according to Eqs. 8.9 and 8.9 (see page 108). For comparison the CE of the parabolic collector of the *2-Theta* for the angles 62–80° (*square*). The dashed line devides the region of classical UAF (*left*) and evanescent SAF light (*right*).

up to 41°, which corresponds to a collection efficiency of ~17% – an almost threefold improvement compared to the ~6% for the prototype objective. See Fig. 4.2 for the relationship between collection efficiency (CE) and NA. The parabolic element can collect fluorescence emitted between 60° and 80°. Approximately 10% of the SAF emitted into these angles is lost due to four bores used for alignment of the inner assembly by interferometry [Fig. 4.3A]. The lower cut-off angle for SAF is selected by means of cylindrical apertures ranging from 62° to 70°, which are inserted into the objective housing [Fig. 4.5]. The inner optics can be used for wide-field imaging within a diffraction-limited 30 μm diameter field of view [Fig. 4.4].

Microscope setup

The microscope objective is used with an inverted Olympus IX71 microscope platform (Tokyo, Japan) using the right side port of the microscope body. A schematic of the optical setup is shown in Fig. 4.6. A circularly polarized beam of a 633 nm, 5 mW helium neon laser (JDSU, Milipitas, CA, USA) is expanded and reduced to a beam waist of about 5 mm (5.3 mm clear aperture of the objective) using an iris in order to prevent excitation through parabolic collector. A dichroic beamsplitter ~15 cm from the side port separates the excitation and emission. In the detection path, a lens (f_1=200 mm) positioned 400 mm behind the objective produces a 1:1 image of the

CHAPTER 4. OPTICAL SETUPS

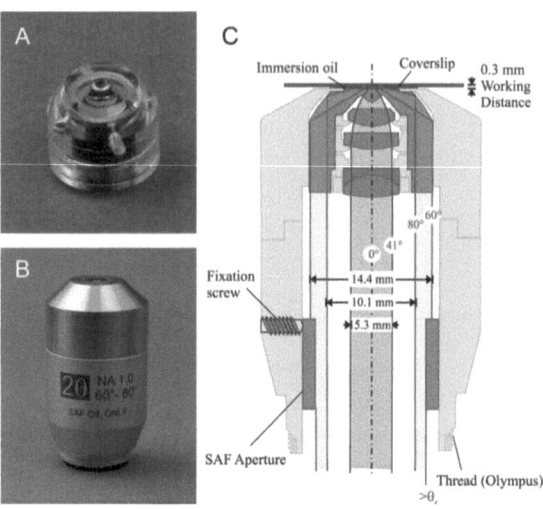

Figure 4.3: The 2-*Theta* microscope objective. (*A and B*) Photographs of the objective and housing. (C) Schematic of the objective. Figure from Ref.[70] © 2011 by the Optical Society of America.

objective's output another 400 mm behind it. There, a lens (f_2=200 mm) collimates SAF and UAF beams. A 45° rod mirror fixed on a glass window reflects UAF and allows SAF to pass through. The signals are each focused onto the 180 µm diameter active area of two identical SPADs (PerkinElmer) positioned in the respective image planes. Spectral filters are employed in the excitation (BrightLine HC 632/22, Semrock, USA) and the detection path (BrightLine HC 676/29, Semrock). A motorized microscope stage (ScanIm 120 × 100, Märzhäuser, Germany) is used for sample scanning. Photon counting was first performed with a FIFO buffered counting card PCI-6602 (National Instruments, Austin, TX, USA) and a customized software programmed with Borland C++ builder. Lateron a transition was made to the more advanced counting card PCIe-6320 (National Instruments) with the possibility of advanced synchronized counting on a 100 MHz time-base. Fast synchronized counting was essential for the results produced in section 5.2.

Microscope alignment procedure

The microscope alignment was achieved performing the following steps:

1. Lateral alignment of the excitation beam: Rotationally symmetric coupling of the laser beam

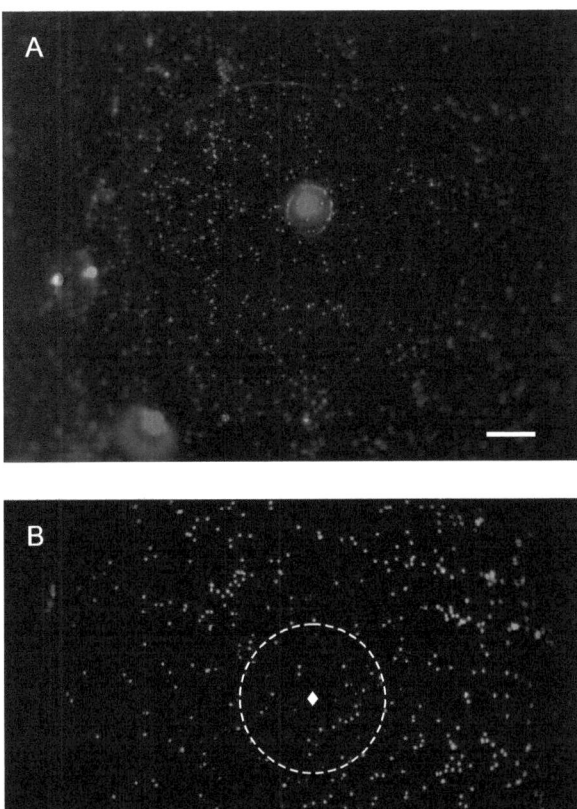

Figure 4.4: Wide-field image of multicolor fluorescent beads (Invitrogen, size range: 0.02-10 μm, colors: blue, green, red) adsorbed to a coverslip from solution. The images were taken using a Sony DXC-390P camera (Tokyo, Japan) on the left side-port of the Olympus IX-71 inverted microscope. (A) Wide-field image without blocking the excitation/emission through the parabolic collector. (B) Wide-field image after blocking excitation/emission through the parabola by placing a circular aperture close to the base of the parabolic collector. The *dashed circle* corresponds to the 30 μm diameter field of view and the *diamond* to the geometrical focus of the parabola. The geometrical focus of the parabola was localized as the point in the field of view where a fluorescent object produces a concentric SAF ring on the camera as in (B). Scale bars=10 μm.

CHAPTER 4. OPTICAL SETUPS

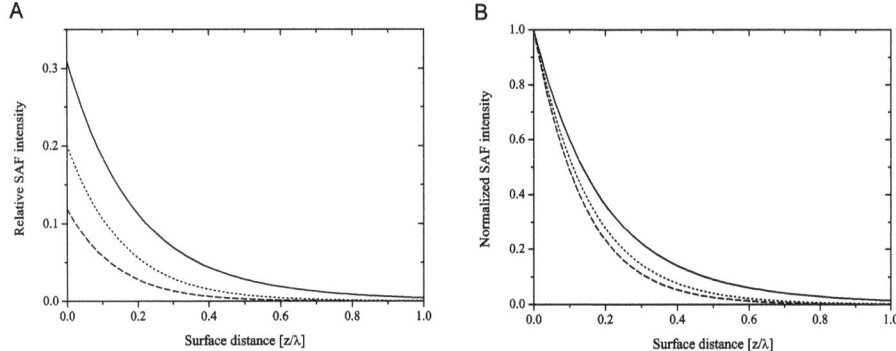

Figure 4.5: Absolute (A) and normalized (B) z-dependence of the fluorescence collection efficiency of the parabolic collector for the angular cut-off at 62° (*solid line*), 66° (*short dashed line*), and 70° (*long dashed line*) for water (n=1.333, θ_c=61°). The normalized curves show the effect of increased surface selectivity with increasing cut-off angle which is, however, obtained at the expense of fluorescence signal.

to the objective was achieved by moving either one or both coupling mirrors until the excitation exiting the objective from above appeared to be rotationally symmetric.

2. Lateral positioning of the excitation beam onto the axis of symmetry of the parabola: This was achieved by applying a layer of autofluorescent red edding and using both coupling mirrors. When the focus of the inner optics is on the symmetry axis of the parabola, the fluorescence intensity image SAF was circular when defocused [Fig. 4.7] and could be focused to a spot smaller than the photosensitive area of 180 µm of the SPAD [Fig. 4.8A]. A lateral displacement of the excitation beam would have lead to loops and knots in the intensity distribution of the refocused SAF. Steps one and two were repeated iteratively until both conditions were met.

3. Axial positioning of the excitation spot onto the focus of the parabola: Unfortunately, the position of the geometrical focus of the parabolic collector lay higher than the focus of the inner optics and could be observed by the fact that SAF und UAF could not be refocused at the same time. This problem required divergent rather than collimated excitation. Correct divergence of the excitation beam obtained using the expander resulted in the coincidence of the two foci in the focal plane allowing for SAF and UAF emission to be refocused at the same time [Fig. 4.8].

4. Alignment of the SPADs: The approximate image plane for the detectors was found using a layer of autofluorescent red edding with a CCD-camera. The respective focal planes were where the refocused fluorescence produced the smallest spot on the camera. A coarse lateral alignment of the detectors was done by eye. The focal spot of excitation scattering (emission filters were removed) was moved onto the photosensitive area of the SPAD with xy-mirrors. The precise lateral alignment was found using a fairly concentrated solution of fluorophores (>100 nM) and a plasma-treated coverslip to prevent adsorption. The respective xy-mirrors were moved until the maximum signal was obtained. Axial alignment was done by line-scanning fluorescent nanobeads adsorbed to the coverslip and moving the respective lens in front of the detector along the optical axis until they showed the narrowest lateral extension.

The alignment steps 2–4 could be performed using an ordinary black and white board CCD-camera (S/W A1-Pro, Conrad, Hirschau, Germany)

Positioning of the objective inside the housing

Initial FCS measurements gave a substantially larger effective volume for SAF than theoretically expected. This could be attributed to improper alignment of the SAF aperture with respect to the output of the parabolic collector. This could be seen readily from the fluorescence image of the output of the parabola [Fig. 4.9A]. As a a result of the misalignment, a small portion of UAF bled into the SAF channel increasing the effective volume V_{eff}. The objective which is glued into the upper component of the objective housing was slightly tilted and was therefore unglued and repositioned [Fig. 4.9B].

Point-spread function

The performance of the objective was investigated in terms of its lateral PSF by sample-scanning 36 nm diameter red fluorescent beads adsorbed from an aqueous solution to the coverslip with the 62° SAF cut-off [Fig. 4.10]. The measured UAF image shown in Fig. 4.10A is in outstanding agreement with the theoritcal focal intensity distribution of a diffraction-limited aplanatic objective with NA 1.0 [Fig. 4.10B]. The lateral SAF-PSF is nearly identical to the UAF-PSF, as it essentially reflects the laser intensity distribution at the surface. The SAF intensity of a bead is around 1.7-fold higher than the UAF intensity. This would be expected from the theoretical collection efficiencies of ~28% for SAF and ~17% for UAF on the surface. In comparison to the prototype system, the lateral resolution was improved from ~430 nm to ~370 nm ($1/e^2$ intensity).

Single-molecule imaging

Figure 4.11 shows SAF and UAF images of single Atto647N molecules coupled to IgG adsorbed to the coverslip. It can be seen from their bleaching/blinking behaviour that they are in fact single molecules. For single Atto647N molecules a signal-to-background ratio of over 20 could be obtained (see also section 5.1.2).

Output of the parabolic collector

Figure 4.12 shows a fluorescence image of the output of the parabola imaged on a Lumenera Infinity CCD-camera. The sample was Dy647 (Dyomics) labelled BSA (Sigma-Aldrich) adsorbed to the coverslip at its isoelectric point in 20 mM citric acid buffer, pH 4.75. At the isolectric point, BSA is known to form a protein monolayer.[22]

CHAPTER 4. OPTICAL SETUPS

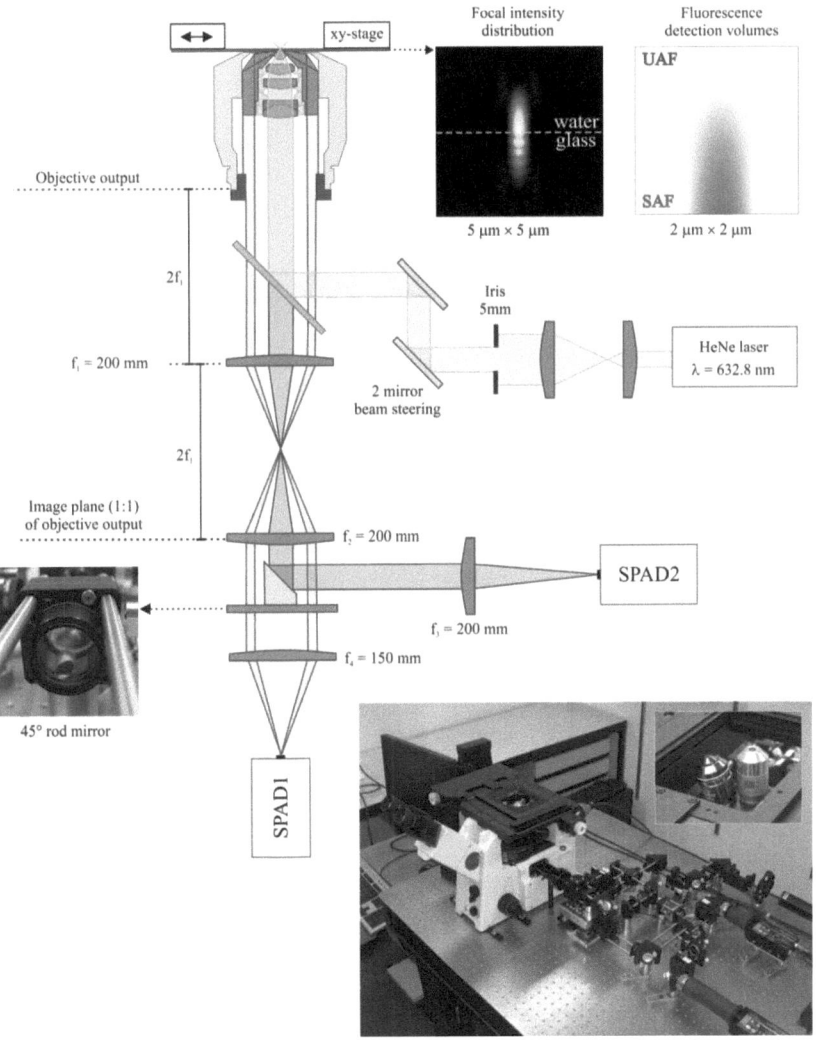

Figure 4.6: Simplified schmatic and photos of the second generation *2-Theta* microscope. Figure adapted from Ref.[70]

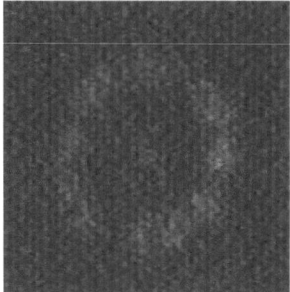

Figure 4.7: Fluorescence intensity image of the defocused SAF in the image plane for correct positioning of the excitation beam onto the parabolic axis; the observed SAF ring is close to circular.

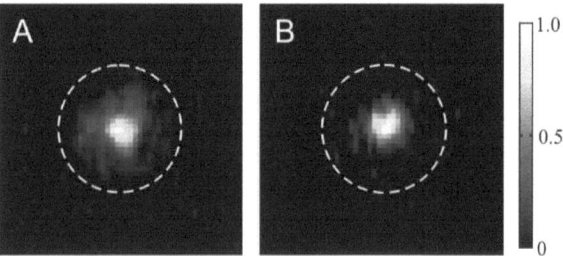

Figure 4.8: (*A and B*) Intensity distributions observed in the image planes of SAF and UAF, respectively. The *dashed circle* delineates the 180 µm active area of the SPAD.

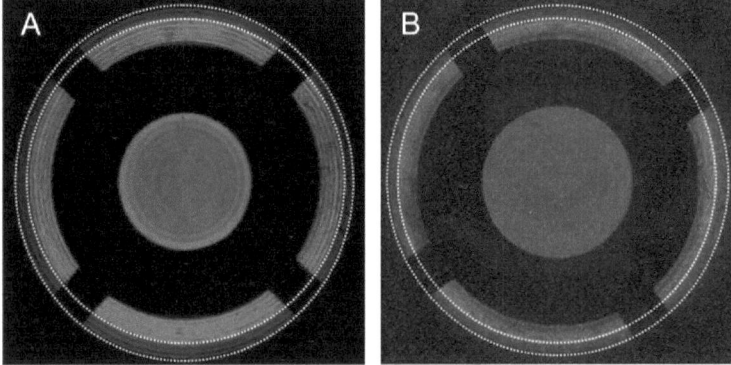

Figure 4.9: Fluorescence images of the output of the objective on a Sony DXP-390P and Lumenera Infinity (Ottawa, Canada) CCD-camera. The sample was Dy647 (Dyomics, Jena, Germany) labelled BSA (Sigma-Aldrich) adsorbed to a coverslip. To visualize the alignment state of the SAF aperture position, the aperture was chosen to have a too small cut-off angle (66°) than required for the RI of the sample (RI=1.41 → θ_c=68°). (A) Incorrect alignment of the SAF aperture where the critical angle (*inner dotted circle*) and the inner radius of the aperture (*outer dotted circle*) are not concentric. (B) Correct alignment. The four dark segments in the output of the parabola stem from the alignment bores.

CHAPTER 4. OPTICAL SETUPS

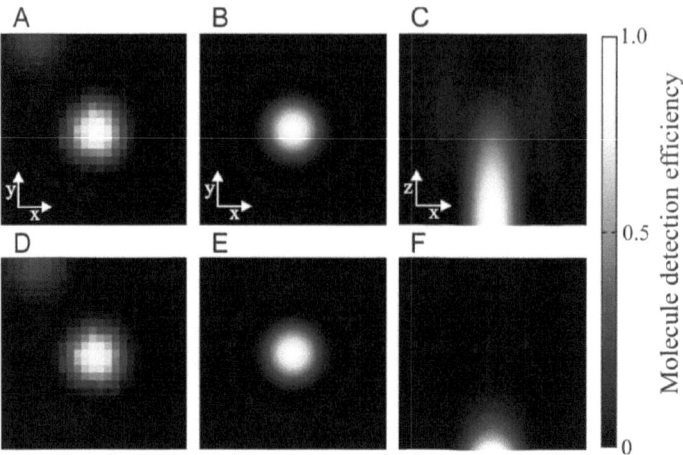

Figure 4.10: Normalized PSFs for UAF collection (*top row*) and for SAF collection (*bottom row*). (A) UAF image of a 36 nm diameter bead. Pixels: 78 nm × 78 nm. (B) Calculated lateral UAF-PSF at the interface and (C) along z. (D) Corresponding SAF image of the bead. (E) Calculated lateral SAF-PSF at the interface and (F) along z. The images have an edge length of 2 μm.

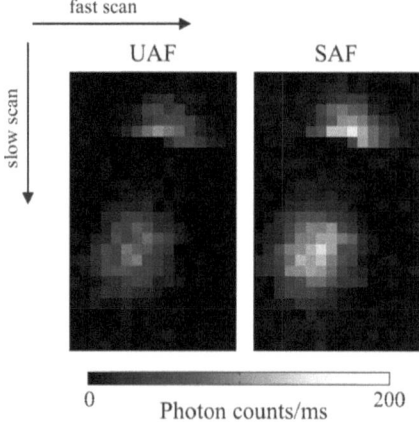

Figure 4.11: Imaging of single Atto647N-IgG molecules adsorbed to the glass coverslip. The molecule above most likely bleached during scanning. The pixels have an edge-length of 78 nm.

32

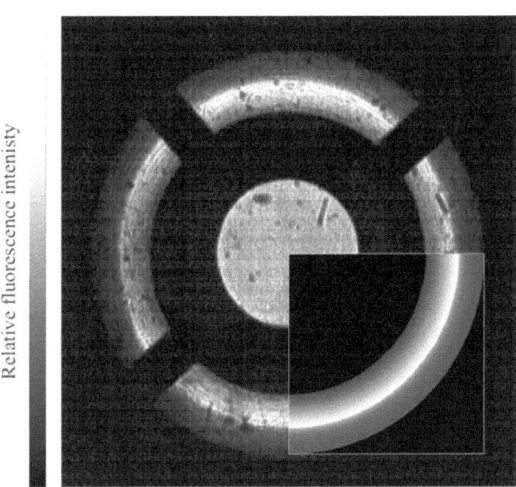

Figure 4.12: Image of the output of the parabolic collector for a monolayer of Dy647-BSA adsorbed onto the coverslip in a glycerol/water mixture with RI=1.41 recorded with a Lumenera Infinity CCD-camera. In comparison, the theoretical ouput shown in the *inset*. The image was taken without a SAF aperture.

4.2 SAF Immunoassay Platform

Disposable test tube

The tube is composed of two polymer components and an O-ring [Fig. 4.13]. The top part is the liquid container. The bottom part containing the optics, concurrently referred to as the substrate, was fabricated by injection moulding of Zeonex® – a cyclo-olefin polymer. An aspheric surface at the bottom side focuses a laser to a homogeneous light disk of 50 µm diameter onto the upper surface of the substrate. The parabolic surface is surrounded by air to produce parallel rays by total internal reflection of the fluorescence emitted at its focal point. The parabola collects SAF in the angular range from 63° to 78° from the optical axis. This corresponds to a collection efficiency of ~27% on the surface [Fig. 4.14] and is as high as for microscope objectives with a NA of over 1.2. The parabolic focal point lies at the center of the upper flat surface – the interface to the analyte solution. All optically crucial interfaces are concealed to prevent damage and contamination.

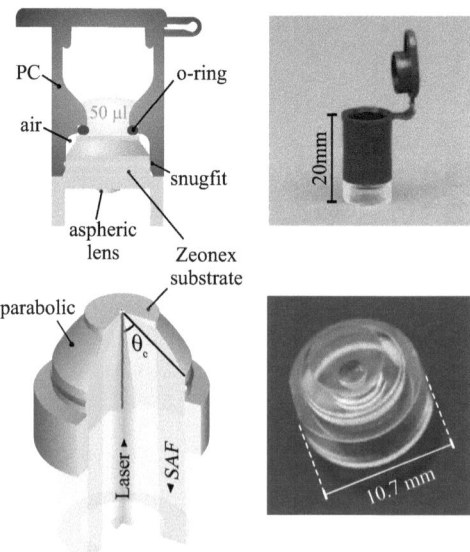

Figure 4.13: Schematic and photographs of the test tube and the optical substrate. Figure from Ref.[71] © 2011 by the Americal Chemical Society.

CHAPTER 4. OPTICAL SETUPS

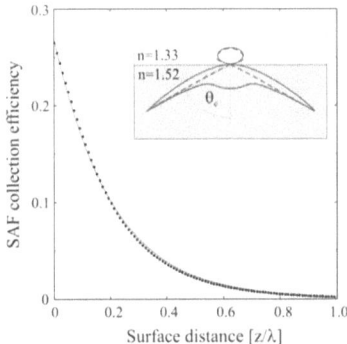

Figure 4.14: SAF emission at the solute/polymer interface. Collection efficiency of the tube for SAF (63°-78°) as a function of surface distance given as a fraction of the emission wavelength λ (*points*). Its z-dependency can be approximated by a mono-exponential with a decay constant of 0.2λ (*solid line*). (*Inset*) The angular distribution of radiation for an emitter with surface distances $z=0$ (*solid*), and $z=\lambda/3$ (*dashed*). Figure from Ref.[71] © 2011 by the Americal Chemical Society.

Fluorescence reader

The instrument is shown in Fig. 4.15. A temperature stabilized 635 nm diode laser is used as excitation source. The laser is aligned by two-mirror beam-steering, and its diameter is reduced to 1.5 mm by means of a circular aperture. A motorized filter wheel is used to swivel a neutral density filter to switch between the excitation intensities of 1 µW and 1 mW. A small reflection prism is used to separate the optical paths of fluorescence excitation and detection. Collimated fluorescence from the tube is refocused with a lens (f=100 mm) through a circular detection aperture of 3 mm diameter, 100 mm behind the lens. The detection aperture serves as a spatial filter to reduce the collected volume element inside the polymer substrate, which is a source of scattered excitation and autofluorescence. The collected volume is thereby reduced to about 70 µm inside the substrate without interfering with the fluorescence collection from the excited 50 µm diameter disk [Fig. 4.16A]. Interference filters are used in the excitation path and in front of the detector. A photomultiplier module is used to detect the fluorescence. A cylindrical cavity on top of the device serves as an adapter for the tube. Due to the design of the substrate optics, the excitation focus is positioned exactly at the center of the substrate simply by inserting the tube into the adapter. The dimensions of the prototype reader are 254 mm × 180 mm × 90 mm. Data acquisition and control was done using incorporated electronic circuits and a laptop via USB.

CHAPTER 4. OPTICAL SETUPS

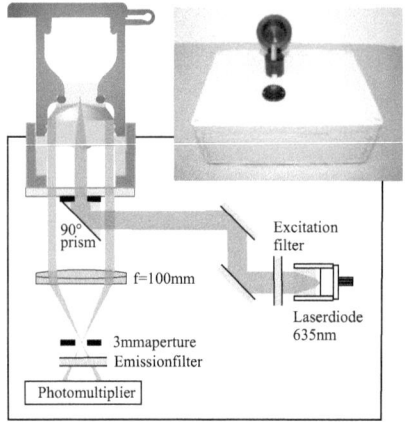
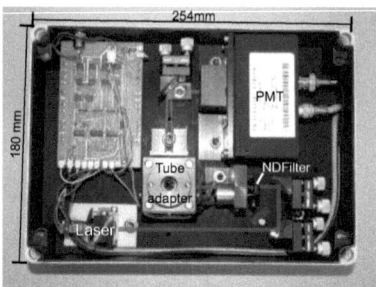

Figure 4.15: Schematics and photographs of the fluorescence reader (*ND*=Neutral density, *PMT*=Photomultiplier). Figure from Ref.[71] © 2011 by the Americal Chemical Society

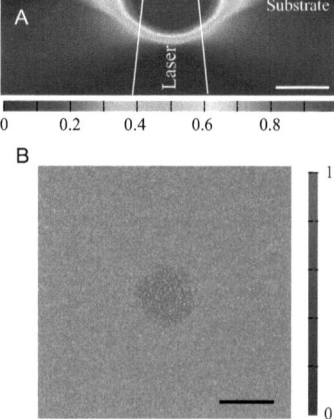

Figure 4.16: (*A*) Spatial collection efficiency function. The collected volume element inside the substrate is reduced to less than 100 µm. The marginal rays of the laser beam are also indicated (*white lines*). (*B*) Fluorescence image of a fluorescence coated substrate after photobleaching on the reader instrument showing the 50 µm diameter laser excitation spot on the surface. Scale bars=50 µm. Figure from Ref.[71] © 2011 by the Americal Chemical Society.

5 Parallel Near- and Far-Field Microscopy*

The angular region for the collection of SAF is covered only by high aperture objectives (NA>1.33) for water/glass. An alternative collection scheme is used, where SAF is collected by means of simple solid parabolic element. The high collection efficiency of this method has even allowed for single-molecule detection.[74] A custom optical configuration is used which allows for the simultaneous, well separated detection of near-field meadiated SAF and the far-field UAF emission modes [Fig. 5.1]. It is based on an inner focusing optics incorporated into a parabolic collector with coinciding focus. The inner optics focuses a laser to a diffraction-limited Gaussian spot at the surface of a coverslip. UAF emission is collected by the focusing inner optics. SAF emission is collected by means of a parabolic element converting into conveniently processable parallel rays. A practical advantage of this approach is that the lowest collected angle lies at the outer margin of the collimated fluorescence ring exiting the parabola. It is technically straightforward to ensure exclusive supercritical angle collection depending on the RI of the sample. The lower limit of collected angles can be easily set to the desired value above the critical angle with a circular aperture. SAF and UAF signals are each detected with single-photon avalanche diodes (SPADs). The separate optical paths for near- and far-field emission generate two laterally completely overlapping fluorescence detection volumes of very different axial extent.

This chapter will cover applications of parallel near- and far-field detection in sub-diffraction microscopy and FCS. These applications were further supported by the development of a second generation microscope system capable of detecting near- and far-field emissiom modes down to the level of single molecules.

---

*The results of this chapter were partially published in:

C.M. Winterflood, T. Ruckstuhl, D. Verdes, S. Seeger. *Phys. Rev. Lett.*, **105**(10), 108103 (2010)[72]

T. Ruckstuhl, D. Verdes, C.M. Winterflood, S. Seeger, *Opt. Express*, **17**(7), 6836 (2011)[70]

C.M. Winterflood, T. Ruckstuhl, N.P. Reynolds, S. Seeger, *ChemPhysChem* **12**(16), 3655 (2012)[73]

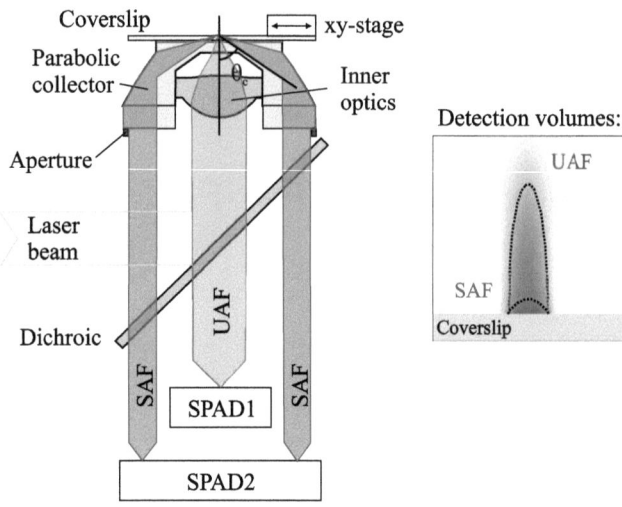

Figure 5.1: (A) Schematic of the optical configuration for parallel near- and far-field microscopy and the two detection volumes where the dotted lines represent the $1/e$-intensity isolines.

5.1 Three-Dimensional Supercritical Angle Fluorescence Microscopy

SAF provides an extremely sensitive measure of the z-position due to its quasi-exponential fall-off along the optical axis. However, the knowledge of the SAF intensity alone does not suffice to determine the axial position of an emitter, as it is governed by its intrinsic brightness. It is, for instance, not possible to distinguish the case where an intrinsically bright emitter is far away from the coverslip, from the case where an intrinsically dark emitter is close to the coverslip. The UAF emitted into surface angles below θ_c is only moderately influenced by an emitter's distance from the interface[75] and can therefore be used to measure its intrinsic brightness. The proportions of SAF and UAF measured completely simultanesously relate directly to the z-position above the coverslip, which in practice can be obtained from the SAF and UAF intensity ratio I_{SAF}/I_{UAF} and the theory of its decay along the optical axis [Fig. 5.2] (refer to page 6 for the properties of fluorescence emission near transparent interfaces). Axial localization requires only knowledge of easily accessible experimental parameters such as RI of the sample, collected angles, and emission wavelength. Similarly to the scheme presented in this work, in differential evanescence nanometry the illumination mode

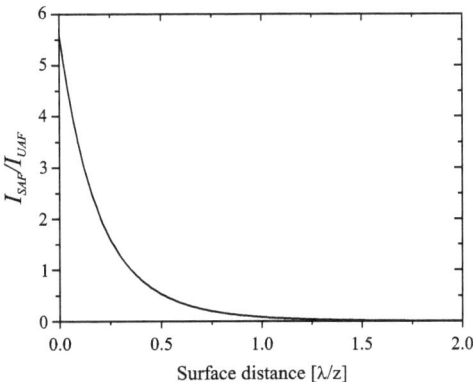

Figure 5.2: Distance dependence of I_{SAF}/I_{UAF} for the prototype near- and far-field microscope. $I_{SAF}/I_{UAF} \approx$ 5.5 for z=0 reflects the more than fivefold higher collection efficiency of the prototype optics for SAF directly at the surface.

is alternatingly switched between TIRF and wide-field.[10] The wide-field image serves to calibrate the TIRF image. In the proposed 3D-SAFM method, the simultaneous detection of SAF and UAF makes it possible to account for fluorescence intensity fluctuations typical for small emitters on any time scale and to measure fast processes.

In principle, it is possible to retrieve absolute emitter-surface distances from I_{SAF}/I_{UAF} and exact knowledge of the relative detection efficiencies which can be obtained by very precise calibration of the microscope. A change of the relative detection efficiencies of SAF and UAF by 1% causes a systematic error of already 2 nm for the absolute surface distance. Frequent recalibration can probably not be avoided as instrumental variations are difficult to keep below 1% over time, e.g. due to aging effects of the detectors. In practice, it is required to measure an object true to scale in all three dimensions, rather than its absolute position relative to the interface. With 3D-SAFM objects can be resolved accurately without the need of very precise knowledge of the position of the interface. This becomes evident if one looks at the z-dependence of I_{SAF}/I_{UAF} which can be fitted well by a mono-exponential function [Fig. 5.3A]. Two point sources differing in their surface distance generate unequal values for I_{SAF}/I_{UAF}. However, the proportionality of their intensity ratios is almost constant for a given axial separation due to the self-similarity of the exponential

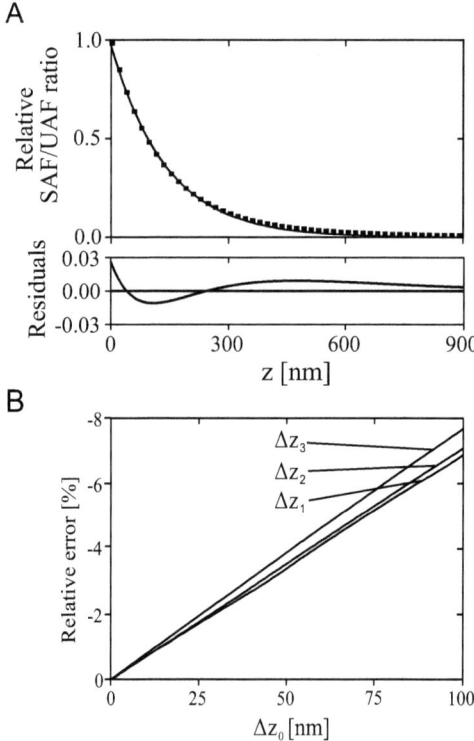

Figure 5.3: Axial localization accuracy of 3D-SAFM. (A) The z-dependency of the relative I_{SAF}/I_{UAF} at a water/glass interface for λ=670 nm fitted by a mono-exponential function (*solid line*). (B) The relative error of the distance measured between two points separated axially by Δz_1=10 nm, Δz_2=50 nm and Δz_3=200 nm as a function of the error Δz_0 in establishing the z-origin. Even for a very large Δz_0 of 100 nm the relative error of the distance measured between two points is below 8%. Figure from Ref.[72] © 2010 by the Americal Physical Society.

function, irrespective of their distance from the interface [Fig. 5.3B] as

$$I_{SAF}/I_{UAF}(z) \approx I_{SAF}/I_{UAF}(z=0) \times e^{-z/d}$$
$$\frac{I_{SAF}/I_{UAF}(z=0) \times e^{-(z+\Delta z)/d}}{I_{SAF}/I_{UAF}(z=0) \times e^{-z/d}} \approx e^{-\Delta z/d} \approx \text{const.},$$
(5.1)

where d is the exponential decay length and Δz ist the axial separation between two points.

5.1.1 3D-SAFM measurements with the prototype microscope

3D-SAFM was first performed with the prototype sample-scanning fluorescence microscope (see section 4.1.1 for a detailed description of the setup). A geometrically well-defined object was used in a first model experiment. In a similar experiment as in Ref.,[76] the z-profile of a silica microsphere of 5 µm diameter was measured [Fig. 5.4]. The bead was coated on its surface with the red-fluorescent dye $DiIC_{18}(5)$, immobilized onto a coverslip and surrounded by an index-matched solution of water/glycerol (see page 97 for sample preparation). Optically speaking, the system was thereby reduced to only the solution/glass interface. The lower SAF collection angle was set to the corresponding critical angle θ_c=69.8° using the annular aperture. The microspheres deliver UAF and SAF images with very different intensity distributions. The depth of the UAF detection volume is roughly the radius of the bead and "sees" its full lateral extent, whereas in the SAF image only the contact region of the particle is visible. The z-profile was calculated pixelwise from the ratio of SAF and UAF intensities according to its z-dependency. The z-profile of the bead is undistorted and follows an ideal sphere with z-values deviating at most by ~60 nm, possibly reflecting the roughness of the particle.

The characterization of the full axial resolution of 3D-SAFM required the use of flourescent point sources. For this, 36 nm diameter red-fluorescent beads were used. The mismatch of their RI with the medium could be neglected as their volume is by far smaller than the extension of the near-field. The beads were measured either in water or in an agarose gel having an identical RI (see page 99 for sample preparation). The low SAF collection angle limit was set to θ_c=61.1°. In one case, the beads were adsorbed from an aqueous solution to the coverslip [Fig. 5.5A-C]. Their normalized SAF and UAF intensity images are indistinguishable, but the SAF intensity is more than fivefold higher due to the higher collection efficiency above θ_c. The indicated z-positions were calculated from the mean value of a bead's image pixels. The z-origin was set to the center of the lowest bead with the highest average I_{SAF}/I_{UAF}. The indicated errors were obtained from the standard error of the mean, which corresponds to the localization accuracy. The variation in the measured z-positions is consistent with the size distribution measured by atomic force microscopy (AFM).[77]

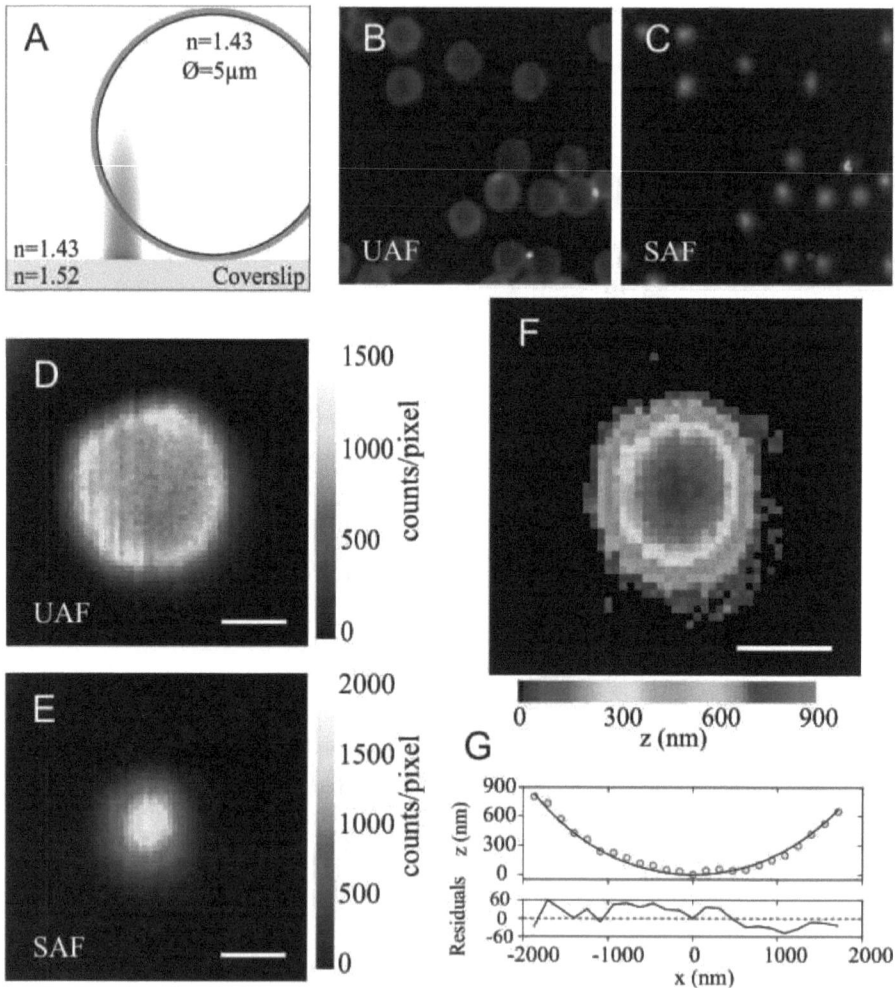

Figure 5.4: 3D-SAFM of a fluorescence coated microsphere. (A) Schematic of the experiment. (B and C) UAF and SAF images of several beads. (D and E) UAF and SAF images of a single bead. (F) 3D-image of the bead surface-contact region for pixels with SAF intensities above a threshold. (G) z-profile through the center of the bead fitted to a circle with 5.1 µm diameter. The sample was scanned with a pixel size of 156 nm and 10 ms integration time. ~1100–2600 counts per pixel were detected in both channels together for determining the z-positions. Scale bars=2 µm. Figure adapted from Ref.[72]

It is to note, that the accuracy in the z-localization is two orders of magnitude higher than the diffraction-limited lateral resolution. The localization accuracy is principally only limited by shot noise ($\sim$60'000 counts per bead on average).

In the second case, the beads were embedded with a heterogeneous z-distribution in a thin film of agarose-gel produced by spin-coating with less than 1 µm thickness [Fig. 5.5D-F]. This was done to avoid having multiple beads in the longer detection volume for UAF preventing a meaningful axial localization. As expected, the accuracy decreased along z as fewer SAF photons are collected. However, all beads could still be localized within an error of 15 nm [Fig. 5.5F].

The method was further extended to cell imaging. 3D-SAFM was tested on the immunostained microtubule network of mouse embryonic NIH 3T3 fibroblast cells. With a diameter of $\sim$60 nm, antibody-stained microtubules represent ideal nanoscale probes and are frequently used for characterizing super-resolution microscopy techniques.[7,54] Figure 5.6 shows the parallel imaging of a cell whose microtubules were stained using a primary antibody towards the subunit α-tubulin and a secondary Cy5-labelled antibody (see page 97 for sample preparation). The high density of microtubles didn't allow to produce a three-dimensional image from I_{SAF}/I_{UAF}. Objects which are laterally overlapping, partly due to the relatively poor resolution of the SAF objective in the xy-plane, cannot be resolved. To overcome this limitation, the cells were treated with the antimitotic agent nocodazole before staining. Nocodazole interferes with the polymerization of microtubules thereby reducing their number. During the process of immunostaining the cell membrane is permeabilized allowing for the antibodies to enter. As a consequence, the interior of the cell is RI-matched with its surrounding buffer. Under a light microscope the cells are virtually invisible. A 3D-SAFM image of a fibroblast cell is shown in Fig. 5.7. The bulk of the network is located within the first wavelength from the coverslip and can be z-localized. More distant microtubules remain elusive to the SAF detection volume and can therefore not be localized and yet are imaged by UAF. At the crossings of laterally overlapping microtubules the calculated z-position represents an average value.

5.1.2 Axial localization with the *2-Theta* microscope

The results shown so far were generated on the prototype microscope system described in section 4.1.1. The newly developed *2-Theta* objective and microscope system presented in section 4.1.2 offers single-molecule sensitivity additionally for UAF allowing for the method to be extended to the axial localization of single molecules. Figure 5.8 shows the measurement of Atto647N (ATTO-TEC, Siegen, Germany) labelled mouse IgG (dye/protein ratio of $\sim$1.5) immobilized onto a coverslip by

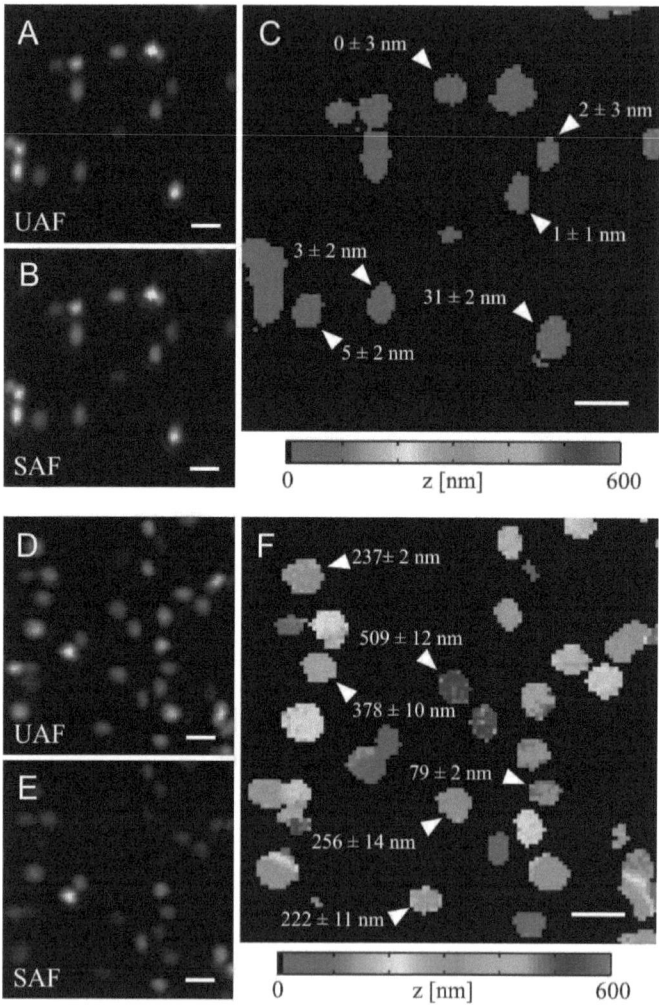

Figure 5.5: 3D-SAFM of 36 nm diameter fluorescent nanospheres. (A) UAF image, (B) SAF image and (C) 3D image of beads adsorbed at the water/coverslip interface. (D) UAF image, (E) SAF image and (F) 3D-image of beads embedded in agarose gel. The sample was scanned with a pixel size of 156 nm and 10 ms integration time. Scale bars=2 µm. Figure from Ref.[72] © 2010 by the Americal Physical Society.

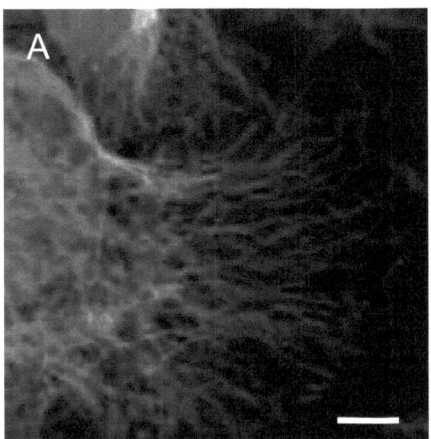 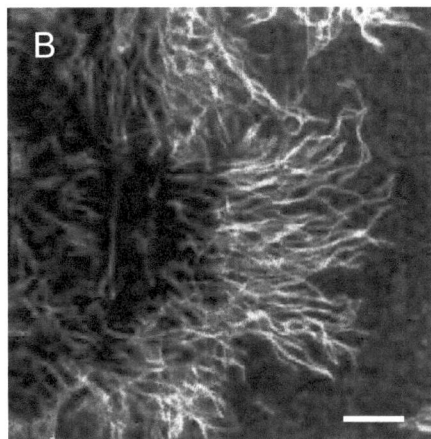

Figure 5.6: UAF (*A*) and SAF (*B*) images of immunostained microtubules of fibroblast cells. Scale bars=10 μm.

non-specific adsorption. The scans were done using circular polarized excitation with 7 μW intensity and 1.5 ms integration time per pixel. The intensity of a molecule in the center of the laser focus was ∼100 kHz for SAF and ∼60 kHz for UAF with a signal-to-background ratio of over 20 for both detection modes. That the spots are indeed single molecules can be seen from their blinking and bleaching behaviour. The broadly similar intensity observed for the molecules was an indication for the free rotation of the antibody-coupled fluorophores on a timescale faster than the fluorescence lifetime. A fixed orientation of the emission dipole moment would result in vastly different I_{SAF}/I_{UAF} [Fig. 5.10]. Fluorophores are excited less efficiently the more their absorption dipole moment points out of the surface plane due to the predominatly parallel polarization of the excitation. This would lead to a selection towards dipoles oriented parallel to the surface and to a lower I_{SAF}/I_{UAF} for incomplete rotational averaging during the fluorescence-lifetime. The assumption of a free rotation of Atto647N is further confirmed by the experiment shown in Fig. 5.9, in which a small aggregate of Atto647N-IgG molecules is bleached step-wise and I_{SAF}/I_{UAF} remains to a large extent constant. For single molecules the error of localization is about a factor of three worse than for nanobeads, which are about ten times brighter. As to negative z-positions, shot-noise can lead to the observation of I_{SAF}/I_{UAF} which is higher than theoretically possible for surface bound emitters. In this case, the z-positions were obtained by linearly extrapolating the z-dependence of I_{SAF}/I_{UAF} to negative z-values using the slope at z=0.

A further benefit of the increased collection efficiency for UAF is the ability to measure axial posi-

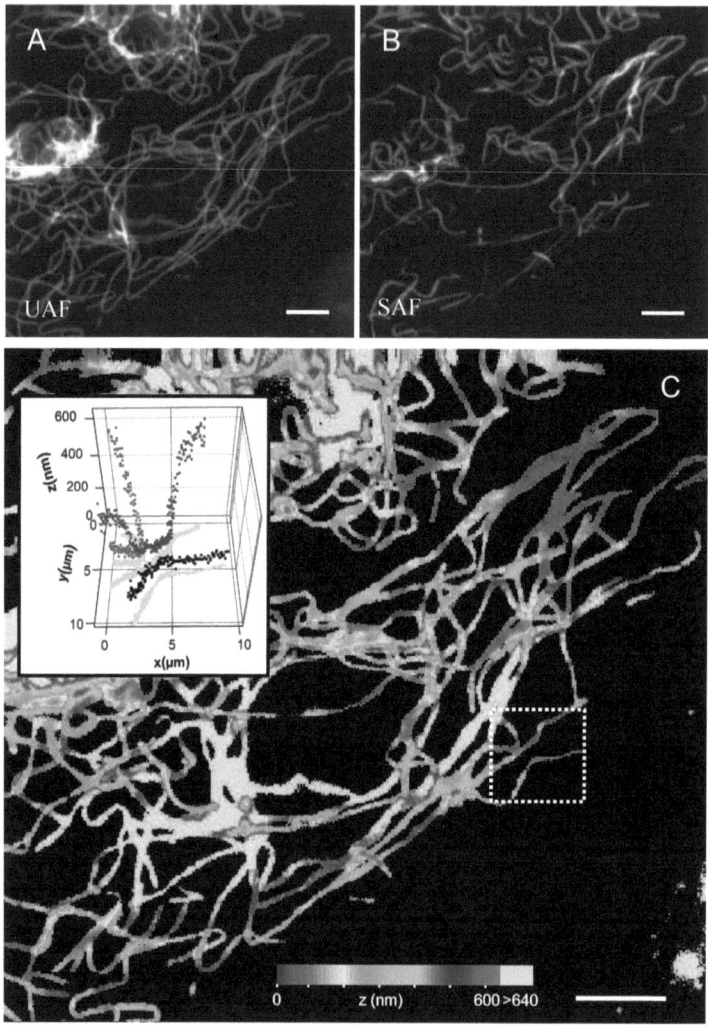

Figure 5.7: 3D-SAFM of the microtubule network of a mouse embryonic fibroblast cell. (*A and B*) UAF and SAF images. (C) 3D-image, where microtubules further away than 640 nm from the coverslip surface (corresponding to <2% of the normalized I_{SAF}/I_{UAF}) were superimposed from the UAF image and are displayed in *gray*. (*Inset*) 3D representation of the pixels in the region outlined by the dashed box. The sample was scanned with a pixel size of 156 nm and 3 ms integration time. Scale bars=10 µm. Figure from Ref.[72] © 2010 by the Americal Physical Society.

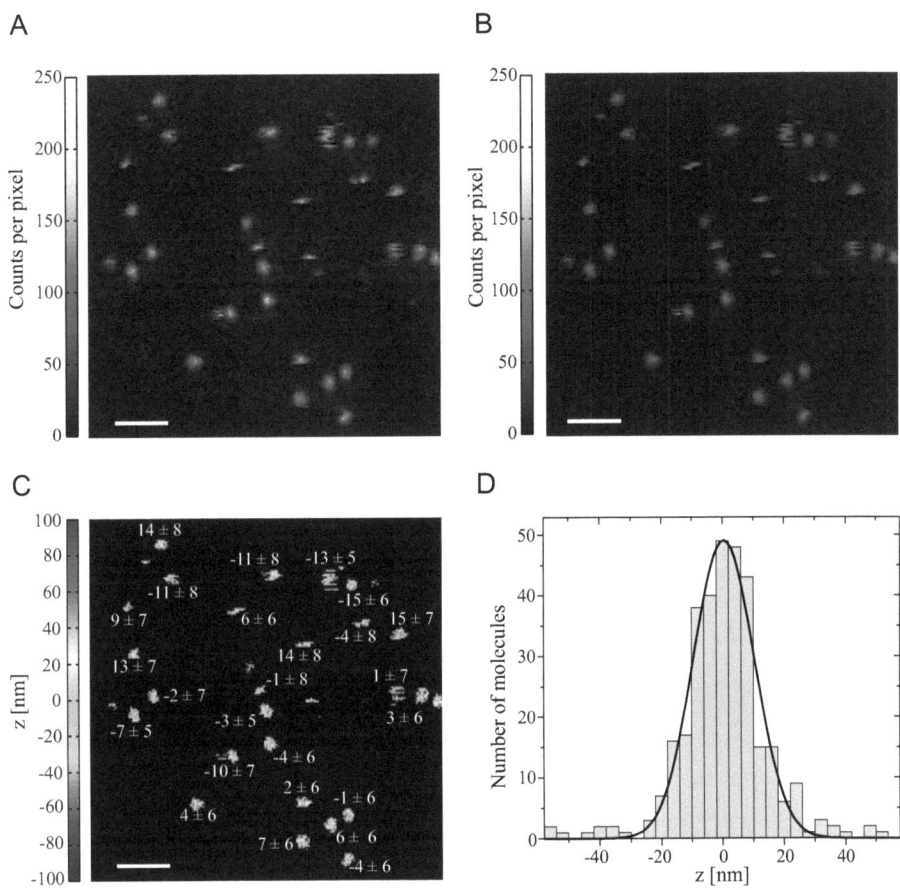

Figure 5.8: Single-molecule imaging and z-localization of IgG-Atto647N immobilized on a coverslip. (A) SAF image and (B) UAF image of a surface area of 13 μm × 13 μm. (C) Calculated z-positions indicated in nanometers. (D) Z-localization histogram of molecules found on a larger surface area of 2000 μm² fitted to a Gaussian distribution (*black line*) with a FWHM=23 nm and standard deviation σ=9.8 nm. Scale bars=2 μm. Figure from Ref.[70] © 2011 by the Optical Society of America.

CHAPTER 5. PARALLEL NEAR- AND FAR-FIELD MICROSCOPY

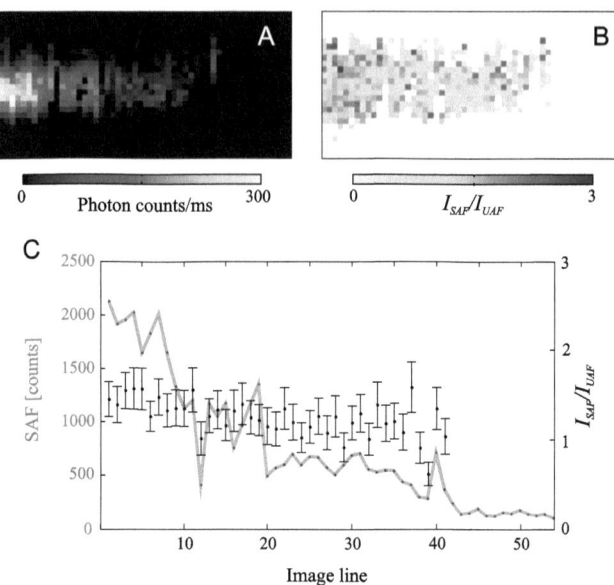

Figure 5.9: Step-wise photobleaching of a small aggregate of Atto647N-IgG molecules adsorbed to the coverslip. (A) SAF line-scan, (B) I_{SAF}/I_{UAF}, and (C) average SAF intensity (*gray line*) and I_{SAF}/I_{UAF} (*black points*) along the lines. Pixelsize: 78 × 78 nm

tions on fast timescales, e.g. of diffusing particles due to improved photon statistics. It is noteworthy that I_{SAF}/I_{UAF} is independent of the lateral position within the laser focus and can therefore be used to track the z-position of particles diffusing close enough to the coverslip for them to emit SAF. In Fig. 5.11 the intensity tracks of SAF and UAF are shown for two bursts produced by single 36 nm diameter fluorescent beads diffusing in a solution of 200 mM sodium chloride. A 62° SAF cut-off and ∼1 µW excitation intensity was used. The bursts in SAF and UAF are different in nature due to the axial movement of the bead. If the bead diffuses further than one emission wavelength from the surface, the SAF count rate remains on the background level despite the high brightness of the particle observed with UAF. When close to the surface, I_{SAF}/I_{UAF} for the beads approached a value of 1.6. The z-trajectory was calculated for time-bins where both intensities exceeded a threshold. Each data point represents an average z-value over the 0.2 ms time-bin. The errors were obtained by error propagation for the division of the shot-noise governed count rates. If the beads were measured in pure H_2O with a low ionic strength, the beads never approached the surface more than

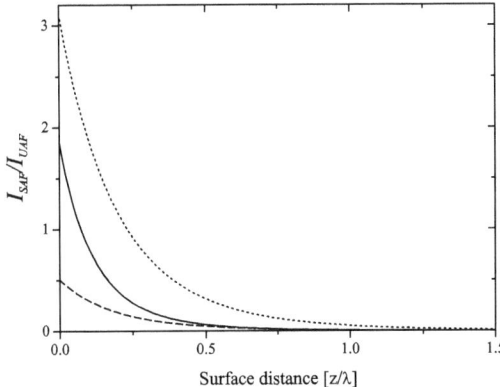

Figure 5.10: Orientation dependence of I_{SAF}/I_{UAF} for the *2-Theta* objective. The z-dependence of I_{SAF}/I_{UAF} is shown for a dipole orthogonal (*short dashed line*) and parallel (*long dashed line*) to the surface, as well as for an isotropically oriented dipole (*solid line*) for water-glass.

∼300 nm due to electrostatic repulsion between the negatively charged carboxy-modified beads and the negatively charged glass surface [Fig. 5.12].

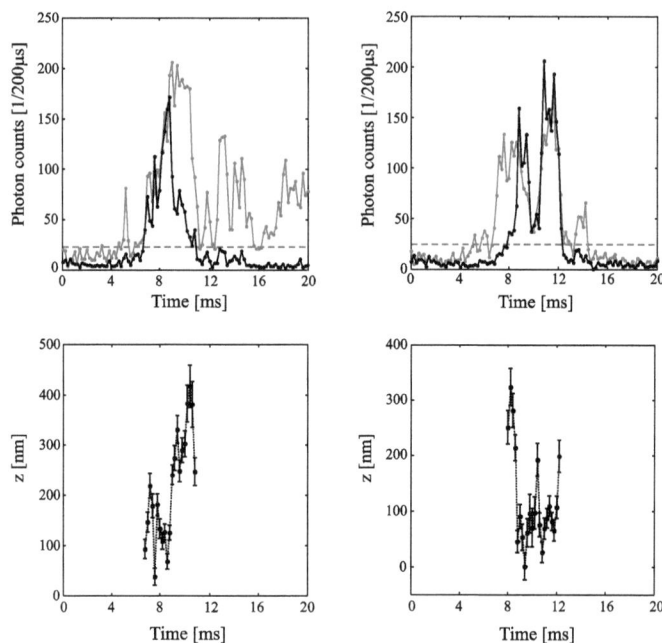

Figure 5.11: Diffusion of 36 nm diamter beads. (*Top row*) SAF (*black*) and UAF (*gray*) bursts for two passages of a bead through the excitation focus in 0.2 ms time bins. (*Bottom row*) Z-positions calculated from the data above for bins where UAF and SAF exceeded the threshold indicated by the *dashed line*. Figure from Ref.[70] © 2011 by the Optical Society of America.

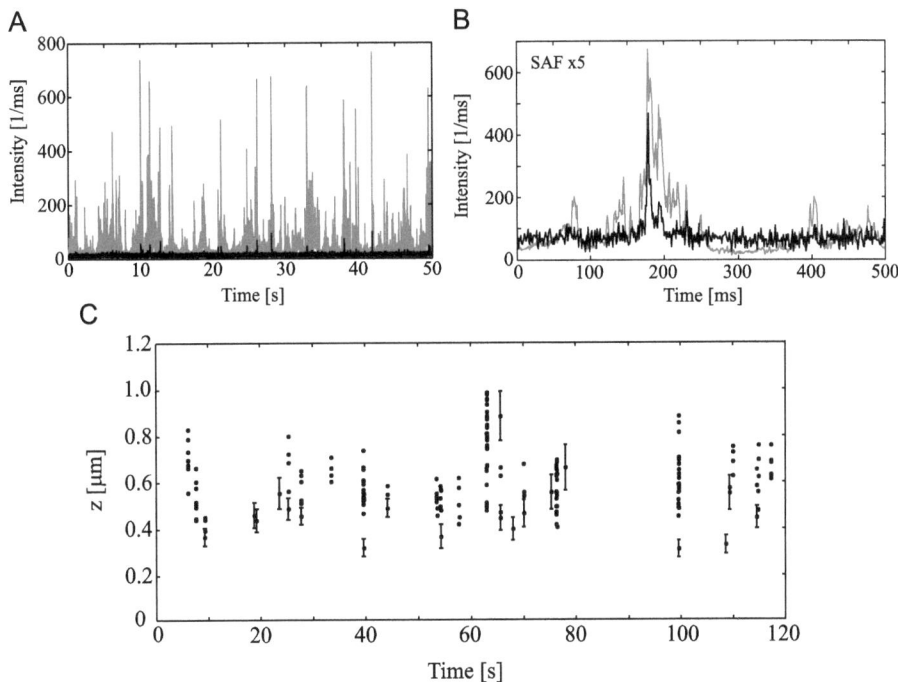

Figure 5.12: Diffusion of 36 nm diameter fluorescent beads in H_2O. (A) A longer intensity track of SAF (*black*) and UAF (*gray*) with almost no significant SAF intensities and (B) a blow up of a region where a bead diffused closer to the surface. (C) Z-positions calculated for the 0.2 ms time-bins where UAF and SAF exceeded a threshold obove the background for a two minute trajectory. A ∼300 nm thick region above the surface is never accessed by the beads due to electrostatic repulsion.

5.2 Parallel Near- and Far-Field FCS

5.2.1 Principles of FCS

Introduction to FCS

FCS is a statistical analysis of the time-dependent fluorescence intensity fluctuations from a small observation volume typically defined by a focused laser beam and a confocal aperture.[78] FCS is usually performed on molecules diffusing in either two or three dimensions. The intensity fluctuations arising from changes in the occupancy number are then subject to a correlation analysis to give the autocorrelation function (ACF)

$$G(\tau) = \frac{\langle I(t)I(t+\tau)\rangle}{\langle I\rangle^2} - 1. \tag{5.2}$$

$I(t)$ is the fluorescence intensity at time t and $I(t+\tau)$ is the intensity at later time $t+\tau$. So the fluorescence signal is correlated with a time-shifted replica of itself for different values of time shift τ. The correlation curve can then be fitted with an appropiate mathematical model, which requires knowledge of the shape of the observation volume. From the fit one can retrieve parameters of interest, such as the fluorophore concentration and diffusion coefficient. In the experimental case of a discrete intensity trace the ACF is calculated as

$$G(\tau_i) = \frac{\frac{1}{M-i}\sum_{n=1}^{M-i} I_n I_{n+i}}{\left(\frac{1}{M}\sum_{n=1}^{M} I_n\right)^2} - 1, \tag{5.3}$$

where M is the number of time-bins with with Δt. Analogously to the autocorrelation of a signal, two signals from two independent detectors can be crosscrorrelated according to

$$G(\tau) = \frac{\langle I_1(t)I_2(t+\tau)\rangle}{\langle I_1\rangle\langle I_2\rangle} - 1, \tag{5.4}$$

with I_1 and I_2 being the signals from two independent detectors.

The straightforward calculation of correlation functions is computationally expensive scaling quadratically with the number of lagtimes τ. As the correlation functions involve time lags that span several orders of magnitude on the time axis, it is more appropriate to evaluate them on a quasi logarithmic scale,[79] which can be calculated according to

$$\tau_j = \begin{cases} 1 & : j = 1 \\ \tau_{j-1} + 2^{\lfloor (j-1)/B \rfloor} & : j > 1, \end{cases} \tag{5.5}$$

with j taking integer values starting with one and running up to some maximum number $j_{max}=n_{casc}B$. B is some integer base number and the sqaured brackets give the integer part of the enclosed expression. The resulting lag times are grouped into n_{casc} cascades with equal spacing of $2^{[j/B]}$.

The autocorrelation function

Here, some properties of the ACF will be discussed. The ACF decays from its initial value with a time-dependence that is determined by molecular diffusion rates. For a molecule in the center of the observation volume, the fluorescence signal at time t and a short time later $t+\tau$ will be very similar (i.e. correlated), as the molecule hasn't had the time to exit the volume. The autocorrelation will decrease with increasing values of τ as the molecules moves out of the volume until the correlation is completely lost, in meaning that the state of the system at time $t+\tau$ has no memory of its state at time t. The time at which the autocorrelation function has decayed to half of its initial value is termed the characteristic diffusion time τ_D and is the average residence time of a fluorophore in the detection volume [Fig. 5.13]. Another interesting property of the ACF is that it provides a measurement of the average number of molecules N in the observed volume, even if the bulk concentration is not known. Provided that the measured fluctuations arise solely from diffusion, the number of molecules is given by the inverse of the intercept at $\tau \to 0$, $G(\tau \to 0) = 1/N$. This correlation amplitude will subsequently be referred to as G_0.

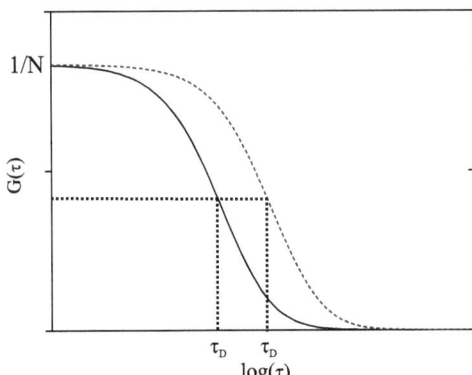

Figure 5.13: Simulated FCS curves for two different diffusion coefficients $D=10$ µm^{2}s^{-1} (*solid curve*) and $D=200$ µm^{2}s^{-1} (*dashed curve*).

Effective volume

The bulk fluorophore concentration C relates to the average number of molecules N in the detection volume through the effective volume defined as

$$V_{\text{eff}} \equiv \frac{N}{C} = \frac{1}{G_0 C}. \tag{5.6}$$

Because the detection volume does not have sharp boundaries it is defined in this unusual way. The value of V_{eff} is the volume containing N fluorophores at a known concentration. V_{eff} can also be calculated from the observation volume spatial profile $W(r)$[80] according to

$$V_{\text{eff}} = \left[\int W(r) dr\right]^2 \times \left[\int W^2(r) dr\right]^{-1}. \tag{5.7}$$

In the case of a confocal volume as described by Eq. 5.17 placed at the coverslip-surface, the experimentally determined V_{eff} for diffusion in three dimensions is still given by the relationship in Eq. 5.6. There is no need to introduce the factor $1/2$ to account for the half-ellipsoid detection volume, as sometimes found in literature.[13,81] Due to the bilateral symmetry of the detection volume (xy mirror-plane), together with the assumption of a completely reflective boundary, the time-dependence of the ACF is identical to the case of a focus placed completely in solution.

Practical considerations in FCS

Triplet kinetics Diffusion is not the only mechnism leading to intensity fluctuations in FCS. Among other processes, intensity fluctuations can also arise from photophysics. For a majority of fluorophores a fraction of illuminated molecules are excited to a triplet state and do not emit photons for a characteristic relaxation time τ_t which is typically on the order of a few tens to hundreds of microseconds. Less commonly the molecules can be excited to a non-fluorescent state due to conformational changes, e.g. cis-trans photoisomerisation. The additional sub-millisecond decay component in the ACF in the case of triplet dynamics [Fig. 5.14] can be accounted for by including the following triplet factor in the ACF[82]

$$G_t(\tau) = G(\tau) \left(1 + \frac{T}{1-T} e^{-\tau/\tau_t}\right), \tag{5.8}$$

where T is the fraction of the molecules in the triplet state and τ_t is the triplet state lifetime. However, superior fluorophores with negligible triplet formation are becoming increasingly available.

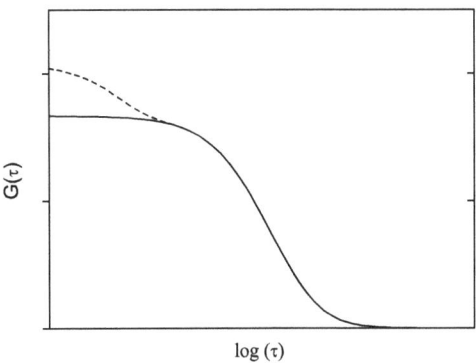

Figure 5.14: Comparison of simulated FCS curves with (*dashed line*) and without (*solid line*) triplet contribution.

Photobleaching and optical saturation At high excitation intensities photobleaching occurs shifting the autocorrelation to shorter times. Another drastic effect on the autocorrelation curve follows optical saturation which can occur already at very low excitation intensities. Optical saturation leads to a flattening of the excitation probability distribution in comparison to the excitation intensity distribution as the fluorophore spends increasingly more time in a state not receptive to a photon (triplet-state, excited-state, dark-state). The detection volume thus becomes apparently enlarged, leading to seemingly longer diffusion times and higher concentrations.[83] In order to avoid effects from excitation intensity, it is common practice to determine the maximum allowed excitation intensity beforehand – the excitation intensity at which there is still a linear relationship between the measured fluorescence intensity and the laser power. To be on the safe side it is better to use lower excitation intensities at the expense of a longer measurement times. The saturation curves for the *2-Theta* microscope for the fluorescent dye Atto655 are shown in Fig. 5.15.

Uncorrelated background The autocorrelation function according to Eq. 5.2 assumes that all the measured fluorescence signal *I* originates from diffusing molecules and is subject to fluctuations. In practice, however, light is detected from several sources of uncorrelated background which do not contribute to the shape of the ACF but reduce its amplitude. Sources of uncorrelated background include detector dark-noise, unrejected rayleigh and raman scattering, as well as sample autofluorecence. The ACF can be corrected for non-correlating background according to the

CHAPTER 5. PARALLEL NEAR- AND FAR-FIELD MICROSCOPY

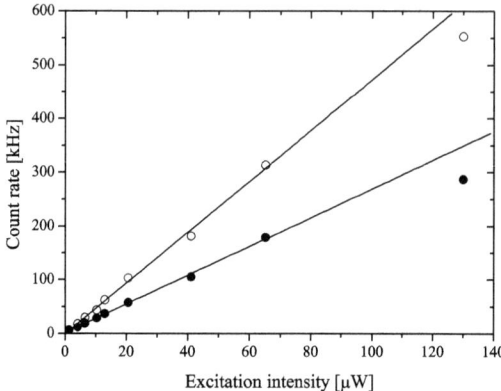

Figure 5.15: Saturation curves of a 10 nM solution of Atto655 for SAF (*closed circles, multiplied by five*) and UAF (*open circles*). The dye was measured in 200 mM NaCl using an oxygen plasma cleaned coverslip and the 66° SAF. The laser excitation intensity was measured at the objective turret with a photometer (Industrial Fiber Optics, Tempe, AZ, USA)

following equation[84]

$$G(\tau)_c = \frac{\langle I \rangle^2}{\langle I - B \rangle^2} G(\tau)_m, \tag{5.9}$$

where $G_c(\tau)$ and $G_m(\tau)$ are the corrected and the measured ACF; I and B are intensity with fluorophores and the background measured in absence of fluorophores, respectively.

Detector afterpulsing and dead-time A common property of single photon avalanche diode detectors typically used in FCS is afterpulsing – the generation of a spurious photon detection event after a genuine photon detection event. The probability for afterpulsing follows an exponential decay extending from hundreds of nanoseconds up to a few microseconds from detecting the real photon. Afterpulses are as such correlated and will lead to a steep increase in the ACF on the latter time-scale. Two common approaches can be used to remove the effects of afterpulsing. The most elegant is to split the emission between two detectors and crosscorrelate their signals according to Eq. 5.4. A second approach is to map the detector afterpulsing probability beforehand with a stable continuous light source.[85] Not having a second SPAD each for SAF and UAF at disposal, in this work, the latter approach was used. The experimental correlation curve $G_c(\tau)$ is then corrected according to

$$G_c(\tau) = G_m(\tau) - G_{ap}(\tau)\frac{\langle F_{ap}\rangle}{\langle F_m\rangle}, \qquad (5.10)$$

where $G_m(\tau)$ are the experimental and $G_{ap}(\tau)$ afterpulsing ACF, F_m and F_{ap} are intensities obtained during the experiment and the measurement of the afterpulsing ACF, respectively.

Another detector artefact in FCS arises from the dead-time of the detector – the time after a photon-count during which the detector is not able to record another event. This can be seen in a drop of the ACF around the dead-time (50 ns for a SPAD). For high count rates, the dead-time effect distorts the correlation curve at small timescales.

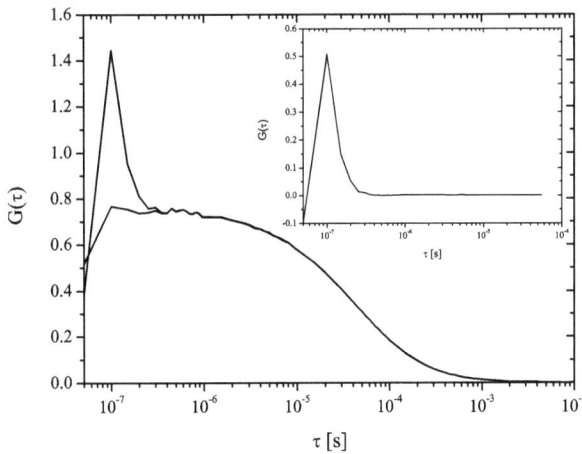

Figure 5.16: FCS curve uncorrected and corrected for detector afterpulsing. (*Inset*) The afterpulsing calibration curve for a SPAD recorded with a white LED lamp.

Count rate per molecule and molecular brightness The count rate per molecule (*cpm*) is a frequently used figure of merit in FCS for comparing instrument performance and state of alignment. *cpm* is the detection rate of photons emitted by a single fluorophore inside the observation volume. A misaglinment or an inherently poor optical performance of the system and the excitation and/or detection are less efficient and the *cpm* will be low/lower. More practical for describing the state of alignment is the molecular brightness (*mB*) – the count rate per molecule and excitation power, given as[86]

$$mB = \frac{countrate}{N \times excitation\ power}. \qquad (5.11)$$

cpm and *mB* are typically measured by FCS. The average number of molecules in the detection volume N is inferred from the amplitude of the ACF either by fitting, or if present, from the plateau at small correlation times. *cpms* have been extracted from bursts for binding events of single molecules inside the observation volume.[13] These are, however, a less robust measure as they depend on the dye's location and trajectory through the focus.

Simulation of correlation curves

The simulations were performed with a routine programmed in C++ with BORLAND BUILDER™ using the numerical library GNU Scientific Library. Simulations were done with continuous position variables and only discretized for read-out on a grid. For SAF-CS a 319 × 319 × 319 15 nm-grid was used, while for UAF-CS a 319 × 319 × 319 60 nm-grid was used. In the case of free diffusion, the size of the box must be chosen to be large enough – much larger than the detection volume (at least one order of magnitude), or else long correlations determined by diffusion in the large volume with low molecule detection efficiency are cut off leading to an ACF which falls off too rapidly.[87] In turn, the amplitude of the ACF is seemingly too large. The laser intensity distribution was computed according to Ref.[88] taking the influence of the water/glass interface into account as described in Refs.[89–91] The detection volume for UAF was computed as the product of the laser intensity distribution and the collection efficiency function calculated by monte-carlo ray-tracing. The detection volume for SAF was calculated as the product of the laser intensity distribution and the relative SAF intensity along z. The collection efficiency of the parabolic element is negligible and was not taken into account. The SAF and UAF detection efficiencies were normalized to unity. N molecules were initialized at random within the box and moved in each coordinate by a random value drawn from a Gaussian distribution with the standard deviation $\sigma=\sqrt{2D\Delta t}$ for every timestep Δt, where D is the diffusion coefficient. Reflective boundary conditions were used; a particle found outside the box was moved back by the value it exceeded the boundary of the box in each dimension. For each timestep Δt and particle the relative detection efficiencies for its position were read and summed over all particles. The detection efficiencies are continuous values and therefore photons statistics was not considered. The resulting "intensity"-track was then auto- and crosscorrelated on a quasi-logarithmic scale of lagtimes τ described by Eq. 5.5.

Specific correlation functions

SAF autocorrelation function for three-dimensional diffusion   The theory describing the SA-FCS autocorrelation curve was adopted from Ref.[15] Under assumption of a Gaussian excitation profile the unnormalized autocorrelation functions for the free diffusion of randomly oriented fluorophores read

$$g(\tau) = g_{xy}(\tau)g_z(\tau)$$

$$g_z(\tau) = \int_{w_a}^{w_b} dw \int_{w_a}^{w_b} dw' f(w)f(w')$$
$$\times \left(\frac{w \operatorname{erfcx}(2\sqrt{D\tau}w') - w' \operatorname{erfcx}(2\sqrt{D\tau}w)}{2(w^2 - w'^2)} \right) \quad (5.12)$$

$$g_{xy}(\tau) = \frac{1}{\pi w_0^2}\left(1 + \frac{4D\tau}{w_0^2}\right)^{-1},$$

where D ist the diffusion coefficient and w_0 the $1/e^2$-radius of the laser focus. The value $w_{a,b} = w(\theta_{a,b})$ is evaluted using the maximum,minimum acceptance angle $\theta_{a,b}$ for SAF collection according to

$$w(\theta) = \frac{2\pi}{\lambda}\sqrt{n_2^2 \sin^2(\theta) - n_1^2}, \quad (5.13)$$

and with

$$f(w_1) = \frac{c(n_1^2 + n_2^2)w_1\sqrt{-n_1^2 + n_2^2 - w_1^{22}}(n_1^2 + w_1^{22})}{3(n_2^2 - n_1^2)^2(n_1^4 + (n_1^2 + n_2^2)w_1^{22})}. \quad (5.14)$$

In a rather heuristic approach the value for G_0 corresponding to the amplitude of the correlation curve at $\tau \to 0$ can be found from a fit to the unnormalized correlation curve [Eq. 5.12] by including a scaling factor γ as a free parameter

$$g'(\tau) = \gamma g(\tau). \quad (5.15)$$

G_0 was then found by evaluating

$$G_0 = \frac{\gamma}{g(\tau)}, \quad (5.16)$$

where $g(\tau')$ is evaluated for a sufficiently small correlation time ($\tau' = 10^{-20}$s was used).

UAF autocorrelation function for three-dimensional diffusion It was not possible to find an appropriate fit-model which could reproduce the experimental UAF ACF of pure diffusion. The commonly used three-dimensional Gaussian model does not hold for the used experimental configuration. The back-aperture of the objective was only slightly underfilled (underfilling fraction β<1.2) producing pronounced diffraction fringes. The large confocal detector aperture (~30 optical units) did not confine the observation volume and reduce the collection from the fringes.[80] Surprisingly, the experimental data could also not be described by the Gauss-Lorentzian model[92] which provides a better description of the axial laser intensity profile. Instead, the UAF ACF was fitted in a very heuristic manner using the three-dimensional Gaussian model for anomalous or subdiffusion, which reads

$$G(\tau) = G_0 \frac{1}{1 + (\tau/\tau_D)^\alpha} \frac{1}{\sqrt{1 + S^2(\tau/\tau_D)^\alpha}}, \tag{5.17}$$

with the diffusion time $\tau_D = w_0^2/4D$, the structure parameter $S = w_0/w_z$, which describes the axial extension of the detection volume and $\alpha < 1$, which describes the degree of anomaly. With $\alpha = 1$, the expression reduces to the standard three-dimensional Gaussian model. A possible explanation for the observation of anomalous diffusion is the change in shape of the laser excitation profile along the optical axis, which becomes manifest in pronounced axial wings.[92]

Autocorrelation functions for two-dimensional diffusion In the case of membrane diffusion, the following two-dimensional Gaussian model applies to both SAF and UAF ACFs

$$G(\tau) = G_0 \frac{1}{1 + (\tau/\tau_D)}. \tag{5.18}$$

The normalized lateral molecule detection functions for UAF and SAF are practically identical for the small distances of the membrane from the coverlip surface still accessible to SAF (less than one emission wavelength).

All FCS fitting software was written in ORIGIN, MATHEMATICA and MATLAB.

5.2.2 FCS simulations and measurements of diffusion in solution

Quantitive results in FCS rely on the size and shape of the detection volume. One common way of calibrating the detection volume parameters is to perform FCS on a species with known diffusion coefficient and concentration.[93] Diffusion measurements were carried out on the red fluorescent dye Atto655 (in its carboxylic acid form, -COOH). The SAF and UAF ACFs were compared with simulations as well as with theoretical models. Atto655 was chosen as it doesn't show any significant triplet-state dynamics and because its diffusion coefficient has already been accurately determined by 2-focus FCS.[94] A difficulty when trying to probe the detection volume for SAF by free diffusion arises from non-specific interaction of the fluorophore with the coverslip glass. This flaws the ACF in that it is shifted to longer times while the amplitude is seemingly decreased. Accordingly, great care needs to be taken for the preparation of the coverslip.

Figure 5.17 shows the parallel FCS measurement with SAF and UAF of the dye at high ionic strength

(200 mM NaCl) to avoid electrostatic repulsion between the negatively charged dye (net charge of -1) and glass[95] (see page 99 for sample preparation). A 63° cut-off was used for SAF. The amplitude (G_0) of the ACF for SAF is over thirty times larger than for UAF. The SAF and UAF ACFs were fitted according to Eqs. 5.12 and 5.17, respectively. The beam-waist w_0 was fixed to the value of 380 nm which was determined by bead-scanning; the diffusion coefficient was fixed at 426 µm²s^{-1}. The average of six separate FCS measurements, each with different lateral positions on the coverslip and newly adjusted focus, gave $V_{\text{eff}} = 144.0 \pm 1.3$ aL for SAF and $V_{\text{eff}} = 5.49 \pm 0.07$ fL, $w_z = 1.04 \pm 0.11$ µm, and $\alpha = 0.76 \pm 0.01$ for UAF. Notably, the relative error for both the SAF and

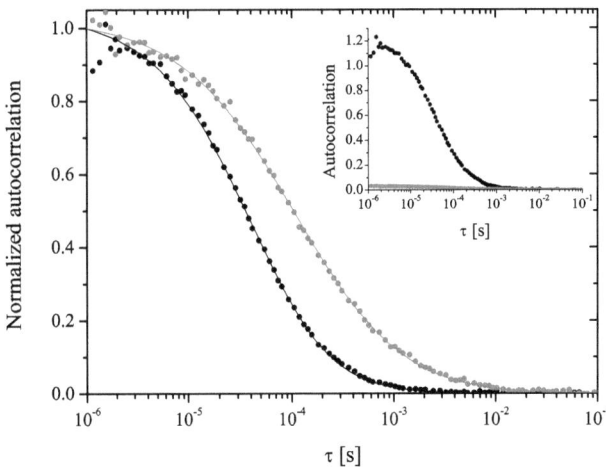

Figure 5.17: Afterpulsing corrected SAF (*black points*) and UAF (*gray points*) ACFs for free diffusion of a 10 nM solution of Atto655 fitted with their respective models (*lines*) and (*inset*) the corresponding unnormalized, background corrected ACFs. Laser power 13 µW, acquisition time 200 s.

UAF effective volumes is only around 1%. The comparison of the experimental ACFs with simulations (see page 60 for the simulation procedure) is shown in Fig. 5.18. Theoretical values for V_{eff} were calculated directly from the observation volume spatial profile according to Eq. 5.7 and gave 136.7 aL for SAF and 6.50 fL for UAF. This is in remarkable agreement with experimental values. With SAF and UAF being measured in parallel it seemed very obvious to evaluate the crosscorrelation functions SAF ⋆ UAF and UAF ⋆ SAF according to Eq. 5.4. As shown in Fig. 5.19 results from simulations match the experimental data very well. There is, however, still little understanding as to the meaning of the crosscorrelation function. It is conceivable that it contains information on

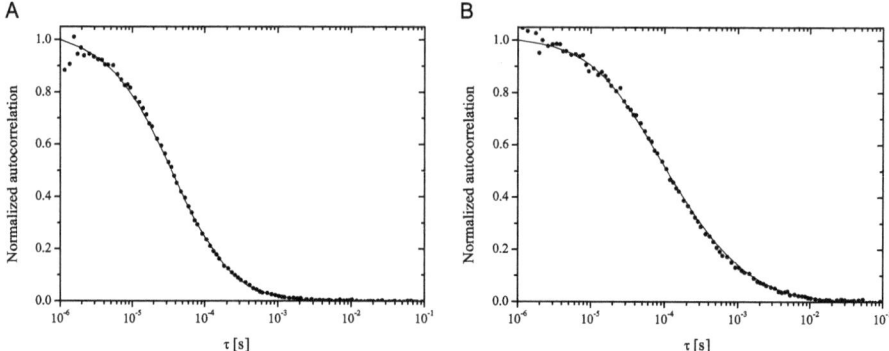

Figure 5.18: Simulated (*lines*) and experimental afterpulsing corrected (*points*) SAF (*A*) and UAF (*B*) ACFs for freely diffusing Atto655.

directional transport along the z-axis or irreversible binding processes.

To further characterize the SAF detection volume, FCS measurements of freely diffusing Atto655 were carried with different SAF cut-off angles, i.e. with different SAF-apertures. Larger SAF cut-off angles of 66° and 70° were used to increase the surface selectivity and thereby further shrink the detection volume. The experimental ACFs are shown in Fig. 5.20 and are in good agreement with the model for SAF-CS [Eq. 5.12]. The experimentally determined effective volumes according to Eq. 5.6 were $V_{eff} = 114.5 \pm 1.0$ aL and $V_{eff} = 127.7 \pm 4.2$ aL for the 66° and the 70° SAF-aperture, respectively. The value for the 66° aperture is in agreement with the value of 112.2 aL calculated using the theoretical observation volume spatial profile according to Eq. 5.7. But the experimental value for the 70° aperture is substantially larger than the theoretical value of 98.1 aL – even larger than compared to the 66° aperture. However, fluorescence collection this far above the fluorescence maximum at the critical angle involves a great loss of fluorescence signal and statistical accurracy and is therefore not practicable.

For freely diffusing Atto655 a *cpm* of 54.4 kHz for SAF and 28.5 kHz for UAF was calculated for a measurement using 67 µW excitation intensity. This corresponds to a *mB* of 8.2 $\times 10^5$ W^{-1} and 4.3 $\times 10^5$ W^{-1} for SAF and UAF, respectively (refer to Eq. 5.11). A 75 µm pinhole was placed in the detection path for UAF for the determining the *mB* and *cpm*.

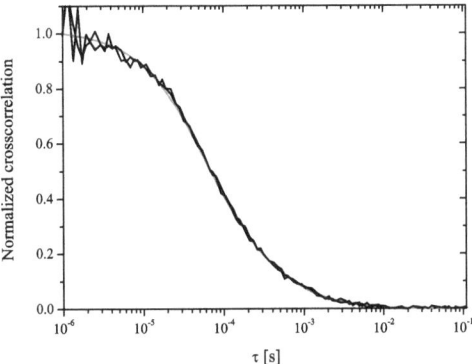

Figure 5.19: Simulated (*gray line*) and experimental crosscorrelation functions SAF ⋆ UAF and UAF ⋆ SAF (*black lines*) for freely diffusing Atto655. Note that the crosscorrelation functions are intrinsically free from afterpulsing effects.

5.2.3 FCS measurements of diffusion in membranes

Biological membranes constitute one of the most integral elements of a cell, defining enclosed spaces or compartments to maintain a chemical or biochemical environment that differs from its surrounding. They can be regarded as a two-dimensional liquid in which their lipid constituents and integral molecules can diffuse more or less freely. An important parameter governing the organization and bioactivity of cell membranes is the lateral mobility of lipids and proteins and the diffusion coefficient in membranes thus represents an important quantity. Several fluorescence techniques are used to measure diffusion coefficients in membranes. Among the most established are fluorescence after photobleaching (FRAP)[96], single particle tracking (SPT)[97] and fluorescence correlation spectroscopy (FCS).[98]

FCS on supported lipid bilayers

The complexity of cellular membranes has motivated the use of simpler model systems. These can be tailored with high precision in size, geometry, and composition for biophysical research. Supported lipid bilayers (SLBs) represent the simplest artifical membranes. They are planar fluid membranes formed by the fusion of unilamellar vesicles onto hydrophilic surfaces. Figure 5.21 shows the SAF and UAF intensity images from a scan of a relatively large area (18 μm × 18 μm) of an SLB with the laterally diffusing, membrane intercalating fluorophore Cellmask™. The SLB was composed of 65mol% DOPC and 35mol% DOPS and was formed on a plasma cleaned coverslip

CHAPTER 5. PARALLEL NEAR- AND FAR-FIELD MICROSCOPY

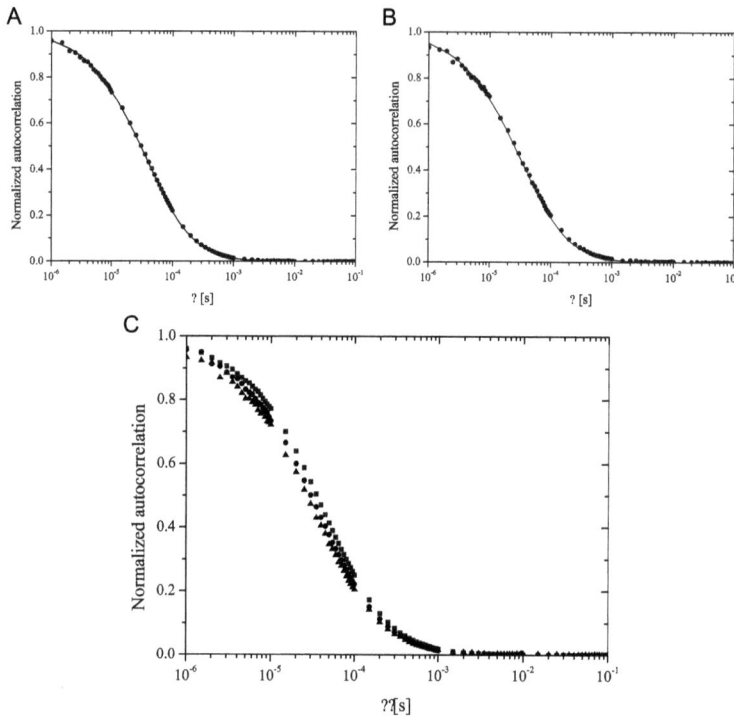

Figure 5.20: (*A and B*) SAF ACFs of freely diffusing Atto655 for a 66° and 70° SAF-aperture, respectively. The *solid lines* are the corresponding fits using a fixed beam-waist of 380 nm and a diffusion coefficient of 426 µm^{2}s^{-1}. (*C*) A comparison between the FCS curves for a 63° (*squares*), 66° (*circles*), and 70° (*triangles*) SAF-aperture showing a faster decay of the ACF with increasing cut-off angle. All curves were corrected for afterpulsing.

(see page 100 for sample preparation). The employed lipid ratio has shown to form a fully mixed lipid bilayer that displays no evidence of phase separation.[99] As can be seen, an intact membrane could be formed on bare glass over a large area. Figure 5.22 shows an FCS measurement of the corresponding membrane. The measurement was performed for 100 s with an excitation intensity of ∼4 µW using a 66° SAF-aperture. Both SAF and UAF ACFs could be described by the two-dimensional Gaussian model [Eq. 5.18]. An independent fit of the ACFs using the beam-waist w_0 as a fixed parameter (w_0=380 nm) yielded a diffusion coefficient of 4.69 ± 0.09 µm^{2}s^{-1} for SAF and 4.72 ± 0.05 µm^{2}s^{-1} for UAF. Triplet dynamics of the Cellmask™ and detector afterpulsing happen

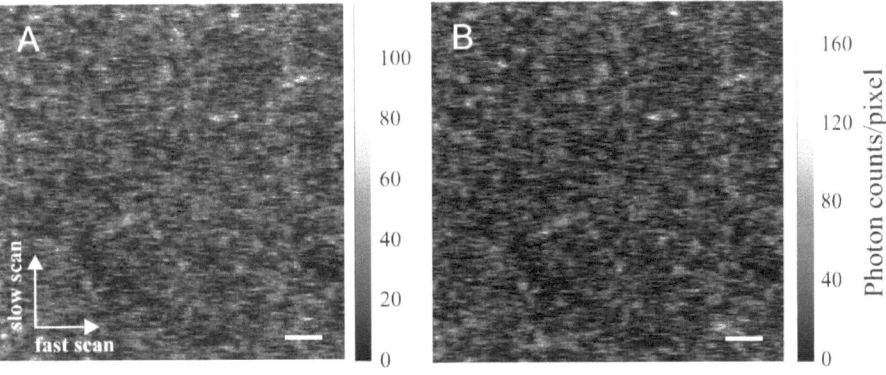

Figure 5.21: (A) SAF and (B) UAF intensity images of a SLB composed of 65mol% DOPC and 35mol% DOPS stained with Cellmask™. The bright horizonal lines arose from the mobility of the fluorophore within the membrane. Scale bars=2 µm.

on a much faster time-scale than the diffusion in the membrane and needn't be accounted for.

Fluorophore orientation in supported lipid bilayers

The question as to the orientation of the membrane probe Cellmask™ in the employed SLB was addressed. For this, I_{SAF}/I_{UAF} was evaluated for singly diffusing molecules in the bilayer. For the studied system (66° SAF-aperture, RI=1.33 of the specimen), I_{SAF}/I_{UAF} of ~1.1 can be expected for an isotropically orienting emission dipole moment at the surface (the water layer between the support and the bilayer is less than one nanometer thick). The results in Fig. 5.23 were indicative of a non-isotropic orientation. The average I_{SAF}/I_{UAF} of around 0.65 was substantially lower than for the isotropic case. This suggested that there was a preferentially parallel orientation of the fluorophore with respect to the membrane. This can be either due to incomplete orientational averaging during the fluorescence-lifetime (because of photo-selection towards parallel oriented fluorophores due to the stronger polarization of the excitation in the parallel plane), or due to a "fixed", more parallel orientation. Studies have shown that fluorophores can be higly oriented in artificial membranes and cell membranes.[100–102] A fixed, more perpendicular orientation would have led to I_{SAF}/I_{UAF} that is higher than for the isotropic case.

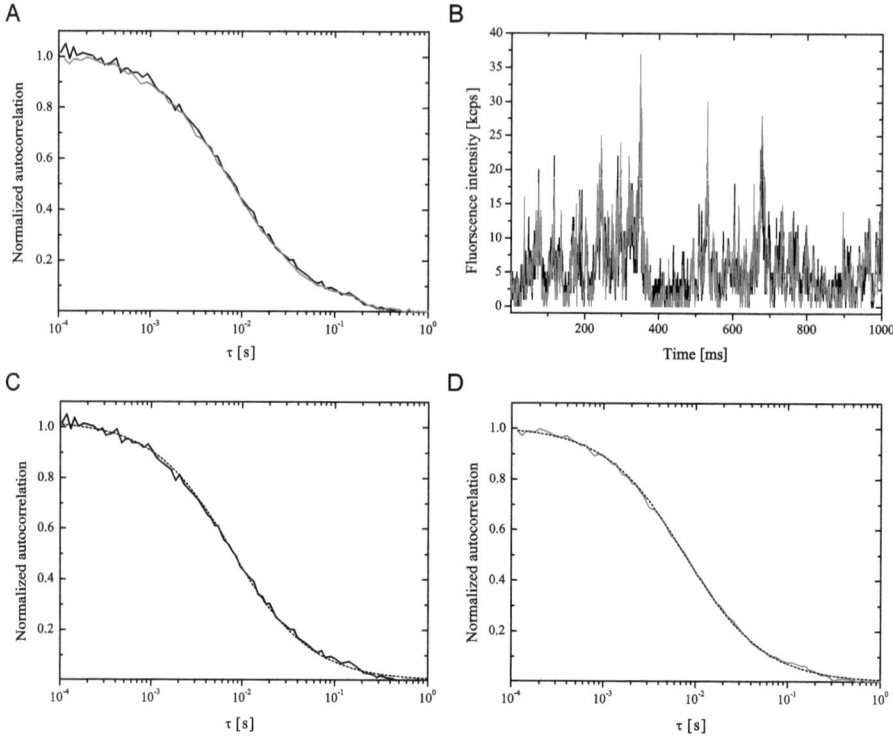

Figure 5.22: SAF- and UAF-CS of the diffusion of the fluorophore Cellmask^{TM} in a SLB. (A) Comparison between the SAF (*black curve*) and UAF (*gray curve*) ACFs and (C and D) their respective fits to a two-dimensional Gaussian model (*dashed*). (B) A one second excerpt from the ms-binned intensity-tracks showing a strong correlation between SAF (*black*) and UAF (*gray*) signal.

Reduction of sample-related artefacts in FCS of membrane diffusion in cells

As opposed to artificial membranes with a well-defined and stable geometry, measuring diffusion coefficients within the plasma membrane of live cells faces additional challenges arising from a mechanical instability or a non-planar topology of the membrane and can lead to wrong assumptions on molecular diffusion if not accounted for. In FRAP,[103] SPT,[104] and FCS[105,106] a membrane curvature has shown to induce an apparent slowing of diffusion without it being anomalous, i.e. the ACF still follows the two-dimensional model for diffusion in a planar membrane, and can therefore go undetected. In FCS, axial movements of the membrane can lead to a distortion of the detection

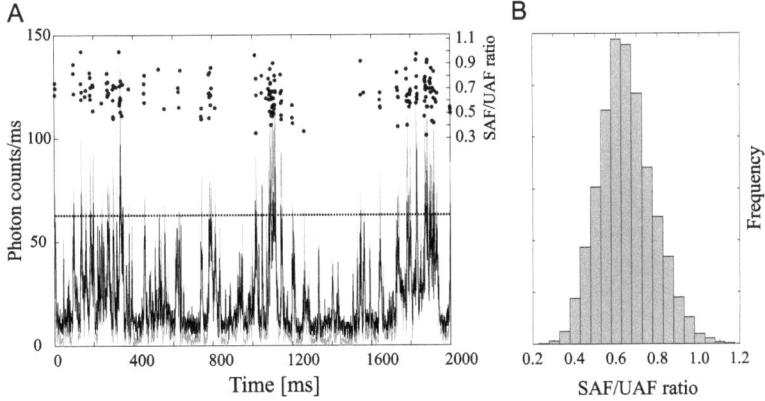

Figure 5.23: (A) A two second intensity-track binned in 1 ms intervals (SAF in black, UAF in gray). I_{SAF}/I_{UAF} (black points) was calculated from bins where the UAF intensity exceeded the threshold of 60 photon counts above the background (dashed line). (B) A histogram of I_{SAF}/I_{UAF} for a time-trace of 100 s. A high excitation intensity of ~20 μW could be used to improve photon statistics as photobleaching does not affect I_{SAF}/I_{UAF}.

area and can even give rise to an additional slow diffusion component.[107] Despite the high precision afforded by modern optical microscopy methods for movements in the lateral plane, the latter techniques suffer from a relatively poor capability for determining spatial differences in the z-axis required to detect a complex topography of the membrane or its axial movements.

In this context a FCS method is presented for determining accurate diffusion coefficients by reducing artefacts introduced by membrane geometry. The technique is highly sensitive towards movements in the z-axis provided by the simultaneous measurement of SAF and UAF emission modes from diffusing molecules. SAF is extremely sensitive towards axial displacements, while UAF is barely influenced by the distance of the fluorophore from the interface within the SAF regime. For a molecule diffusing purely laterally inside a planar and horizontal membrane, UAF and SAF emission intensities will be highly correlated for a transit through the two detection volumes. In the case of a rugged, inclined or axially fluctuating membrane, molecules diffusing within will produce temporally uncorrelated near- and far-field emissions [Fig. 5.24]. The method consists in computing the autocorrelation curves for the simultaneously measured UAF and SAF signals each. The two underlying intensity tracks are subjected to a statistical analysis in terms of their

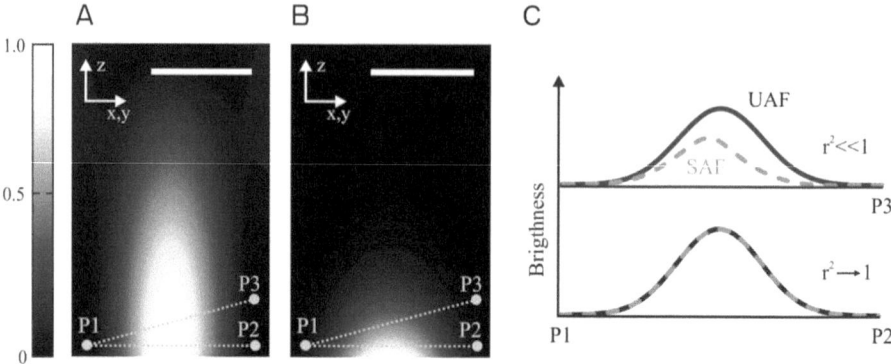

Figure 5.24: Schematic of the principle of parallel SAF- and UAF-CS. If a molecule diffuses through the detection volumes for UAF (A) and SAF (B) purely laterally (*path from P1 to P2*), it will produce highly correlated UAF and SAF signals (*bottom graph in C*). If the trajectory of the molecule through the detection volumes contains an axial component (*path from P1 to P3*), SAF and UAF signals will show a decreased correlation (*top graph in C*) which influences the time-dependence of the SAF and UAF autocorrelation functions. Scale bars=0.5 μm. Figure from Ref.[73] © 2012 by John Wiley & Sons.

degree of correlation using the Pearson's linear correlation coefficient r according to

$$r = \frac{1}{n-1} \sum_{i=1}^{n} \left(\frac{I_i^{SAF} - \langle I^{SAF} \rangle}{\sigma^{SAF}} \right) \left(\frac{I_i^{UAF} - \langle I^{UAF} \rangle}{\sigma^{UAF}} \right), \qquad (5.19)$$

where n denotes the number of time-bins, I^{SAF}, I^{UAF} the intensity of the i-th time-bin, $\langle I^{SAF} \rangle$, $\langle I^{UAF} \rangle$ the mean intensity and , σ^{SAF}, σ^{UAF} the standard deviation of the intensities for SAF and UAF, respectively. The value of r ranges from 0 for completely uncorrelated to 1 for completely correlated signals. r was evaluated for the 1 ms-binned fluorescence intensity tracks.

In a first experiment SLBs were used as a simple model system for the cellular plasma membrane. Because of their planar geometry and stability provided by the solid support they represented the ideal case where contributions of the membrane to the FCS measurement could be largely ruled out. Figure 5.25 shows the measurement of the diffusion of the lipid membrane probe Cellmask™ in the bilayer. The intensity tracks for the simultaneously measured SAF and UAF were highly correlated [Fig. 5.25B]; the Pearson's correlation coefficient r according to Eq. 5.19 was 0.85. The SAF and UAF autocorrelation curves were correspondingly almost identical [Fig. 5.25A]. The autocorrelation curves were fitted to Eq. 5.18 for evaluation of D, yielding 5.01 ± 0.06 μm^{2}s^{-1} and

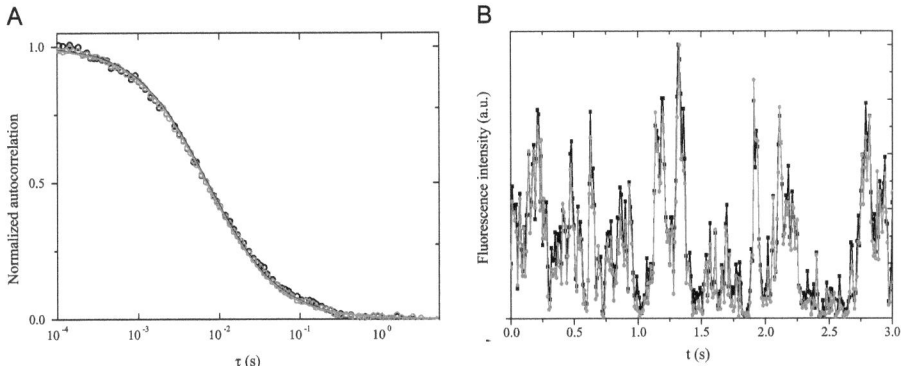

Figure 5.25: Autocorrelation functions for SAF (*black circles*) and UAF (*gray circles*) measured over 300 s and their corresponding fits (*black and gray lines, respectively*). The fitted value for D was 5.01 ± 0.06 µm^{2}s^{-1} and 5.08 ± 0.04 µm^{2}s^{-1} for SAF and UAF, respectively. (B) Typical parallel intensity traces binned in 10 ms intervals. The SAF (*black*) and UAF (*gray*) signals showed a very high correlation (r=0.85). The results were obtained using an excitation intensity of 4 µW. Figure from Ref.[73] © 2012 by John Wiley & Sons.

5.08 ± 0.04 µm^{2}s^{-1} for SAF and UAF, respectively. The relative difference between the determined diffusion coefficients was merely 1.4%. For fitting, w_0 was fixed at 380 nm. Figures 5.26A and B show UAF and SAF images of fibroblast cells with their plasma membrane stained with Cellmask^{TM} (see page 100 for sample preparation). In the SAF image, only the surface-near region of the membrane one wavelength above the coverslip was visualized, revealing positions were the cells adhered to the coverslip. There was a high background from fluorophores adsorbed to the coverslip, but nevertheless the dye partitioned well into the plasma membrane. UAF, on the other hand extends deeper into the specimen (∼1-2 µm) and captured a larger axial section of the cells. The ratio of SAF to UAF intensity is a sensitive measure of the axial position of its source and has been used to determine axial positions of fluorescent emitters and even single molecules with nanometer accuracy (see chapter 5.1). With this powerful tool at hand, ideal positions for FCS measurements were identified from the I_{SAF}/I_{UAF} image [Fig. 5.26]. A high I_{SAF}/I_{UAF} in a cell was indicative of a position where the basal membrane was in contact with the coverslip and where contributions of fluorescence from planes above were minimal , i.e. from the apical membrane or due to separation of the fluorophore into intracellular organelles. Over large regions of the fibroblasts I_{SAF}/I_{UAF} was very low (less than 0.5) and was most probably due to additional fluorescence from the apical membrane. Fibroblast are known for having a very oblate shape with thicknesses of even less than 1 μm.[108] The exact knowledge of the membrane position is a strong benefit, as a vertical mispositioning of the objective focal plane with respect to the membrane can strongly influence the autocorrelation function and therewith assumptions on molecular diffusion. I_{SAF}/I_{UAF} for fluorophores adsorbed to the coverslip was in the order of 1.1 and is in good agreement for an isotropically oriented emission dipole moment. In contrast, the highest ratios within the plasma membrane were 0.6–0.7. The effect of the higher RI of the cytosol compared to the buffer cannot explain this difference. This discrepancy was most probably due the orientation of the dye within the plasma membrane (see page 67). Figure 5.27 shows FCS measurements of the diffusion of Cellmask^{TM} in the plasma membrane. FCS measurements were performed in locations displaying very high values for I_{SAF}/I_{UAF}. Before data collection, immobile fluorophores (e.g. adsorbed to the coverslip) were bleached for several seconds using a high excitation intensity. Figures 5.27A and B show FCS measurements and a portion of the corresponding fluorescence intensity tracks for a position with no or little contribution from membrane geometry (*diamond* in Fig. 5.26C). The SAF and UAF intensity tracks showed a strong correlation (r=0.74). The fit-values of D, 0.456 ± 0.005 μm^{2}s^{-1} and 0.441 ± 0.007 μm^{2}s^{-1} for SAF and UAF, respectively, were within 3.4%. In contrast, Figs. 5.27C and D show an FCS measurement in a position of the membrane (*circle* in Fig. 5.26C) where the SAF and UAF signals showed a low correlation (r=0.59). Both SAF and UAF autocorrelations could be fitted nicely with the two-dimensional diffusion model, though with values for D which differed by 33%. The obtained diffusion coefficients were 0.702 ± 0.007 μm^{2}s^{-1} and 0.473 ± 0.004 μm^{2}s^{-1} for SAF and UAF, respectively. It is obvious that here the FCS curves were influenced by membrane topology, membrane fluctuations or contributions from fluorescence above the membrane and either of the two values for the diffusion coefficient is erroneous. In other words, SAF and UAF autocorrelation curves flawed by contributions of membrane geometry can display a different time-dependence together with a low degree of correlation. It should be stressed that for standard confocal FCS, which is represented by UAF-CS, membrane diffusion overlaid by membrane effects can lead to autocorrelation curves which apparently follow the two-dimensional Gaussian model. Only the simultaneous measurement of SAF allows identifying such spurious FCS curves. The relationship between r-value and the difference in diffusion coefficients for SAF and UAF for a larger set of FCS measurements in an SLB and cell membranes is shown in Fig. 5.28. It is a well known fact that the fluidity of a cell membrane is not constant over time and phase of the cell-cycle,[109] is subject to local heterogeneity[110] and cell-to-cell variability.[111] It was therefore more appropriate to analyze diffusion coefficients in terms of relative rather than absolute values. For the SLB, there was a very small variation (<6%) between the diffusion coefficients determined from the SAF and

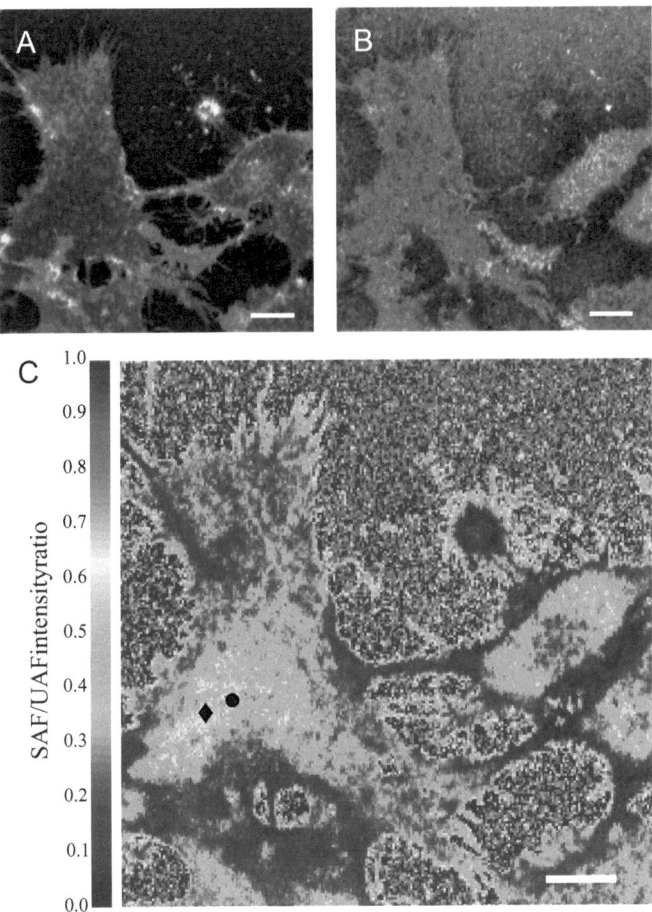

Figure 5.26: Fluorescence intensity scan of fibroblast cells stained with Cellmask[TM]. Images obtained from UAF (A) and SAF (B) in parallel. (C) I_{SAF}/I_{UAF} image after background correction. The sample was scanned with a pixel-size of 312.5 nm and 1.3 ms integration time using an excitation intensity of 1 µW. Scale bars=10 µm. Figure from Ref.[73] © 2012 by John Wiley & Sons.

UAF autocorrelation functions. This is reflected by the consistently high r-values. In the plasma membrane of cells, however, there were very large discrepancies between diffusion coefficients derived from SAF and UAF and r-values. There was a clear trend in that the SAF and UAF diffusion coefficients approached a common value – the more accurate diffusion coefficient – the higher the

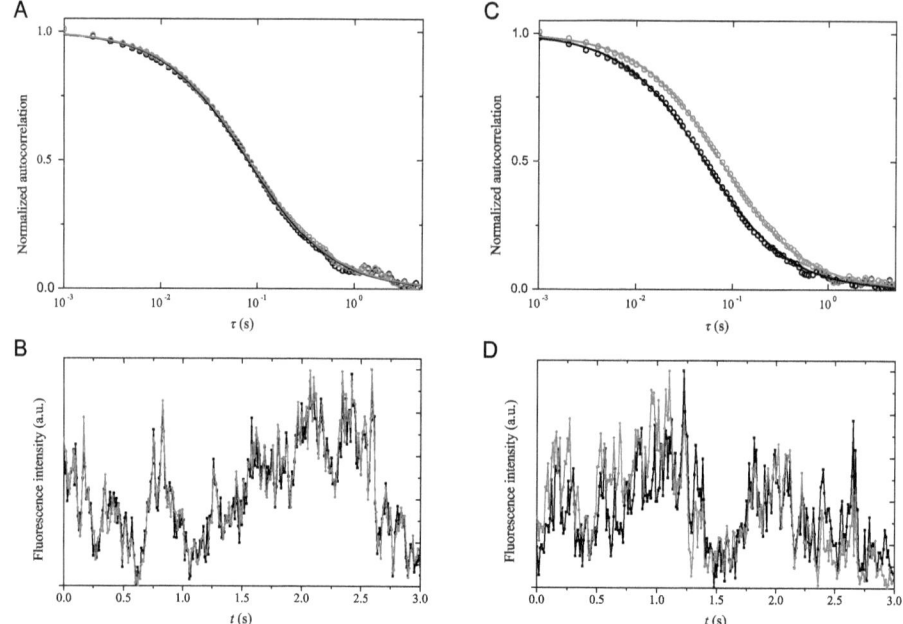

Figure 5.27: Diffusion of Cellmask[TM] in the plasma membrane of fibroblast cells. (A) Autocorrelation functions for SAF (*black circles*) and UAF (*gray circles*) and their corresponding fits (*black and gray lines, respectively*). The fitted value for D was 0.456 ± 0.005 µm²s⁻¹ and 0.441 ± 0.007 µm²s⁻¹ for SAF and UAF, for SAF and UAF, respectively. (B) A representative portion of the intensity track of the FCS-curve in (A) binned in 10 ms intervals showing very similar SAF (*black*) and UAF (*gray*) signals (r=0.74). (C) The fitted values for D were 0.702 ± 0.007 µm²s⁻¹ and 0.473 ± 0.004 µm²s⁻¹ for SAF and UAF, respectively. (D) An excerpt of the intensity track underlying the FCS-curve in (C) binned in 10 ms intervals showing uncorrelated SAF (*black*) and UAF (*gray*) signals (r=0.59). The FCS measurements were performed over 300 s using an excitation intensity of 1 µW. Figure from Ref.[73] © 2012 by John Wiley & Sons.

values of r were. At this point it should be mentioned that a 75 µm diameter pinhole was placed in front of the detector for UAF to reduce the axial extension of the UAF detection volume and to make the comparison with standard confocal FCS justifiable.

FCS measurements of the membrane diffusion of Cellmask[TM] were additionally performed in a second type of cells, namely HeLa cells. This was done to investigate if the problem of membrane geometry is a general issue for determining diffusion coefficients across cell lines. Images of a HeLa

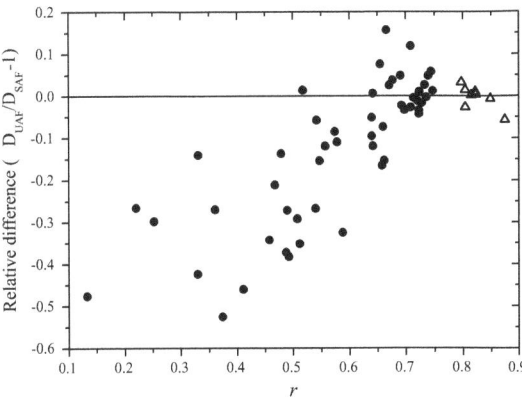

Figure 5.28: Relationship between the relative difference in diffusion the coefficient obtained from the UAF and SAF autocorrelation curves and the correlation value r. *Circles* represent values for the plasma membrane of fibroblast cells, where measurements were taken in different cells and positions. *Triangles* represent measurements in different positions of an SLB. r was calculated from the 1 ms-binned fluorescence intensity tracks. Figure from Ref.[73] © 2012 by John Wiley & Sons.

cell are shown in Fig. 5.29 (see page 100 for sample preparation). Again, FCS measurements were performed in regions with high values for I_{SAF}/I_{UAF}, each over 300 s using an excitation intensity of 1 µW. Five consecutive measurements were performed at the positions indicated in 5.29 C by *black circles*. The Pearson coefficient r of 0.79 ± 0.1 was as high as measured for SLBs. Correspondingly, also the diffusion coefficients obtained from SAF-CS (D=1.34 ± 0.06 µm^{2}s^{-1}) and UAF-CS (D=1.27 ± 0.07 µm^{2}s^{-1}) were very similar. Representative SA- and UAF-FCS curves are shown in Fig. 5.30. These results suggest that in the measured regions, the basal membrane of HeLa cells makes a very flat and stable contact with the coverslip support. Notably, diffusion in HeLa cells is over two times faster than in fibroblasts.

There is some controversy as to the influence of the support on the mobility of lipids in the membrane. It is arguable if measurements of diffusion in the basal cell membrane give the same results as for a free standing portion of the membrane. Studies have shown discrepancies between diffusion coefficients measured in supported lipid bilayers and free standing giant unilamellar vesicles owing to frictional coupling between the solid support and the inner leaflet (inner leaflet, in reference to the vesicular form) as well as coupling between inner and outer leaflet of the bilayer.[112] On the other hand, there are studies which claim that the two leaflets are completely independent[113]

CHAPTER 5. PARALLEL NEAR- AND FAR-FIELD MICROSCOPY

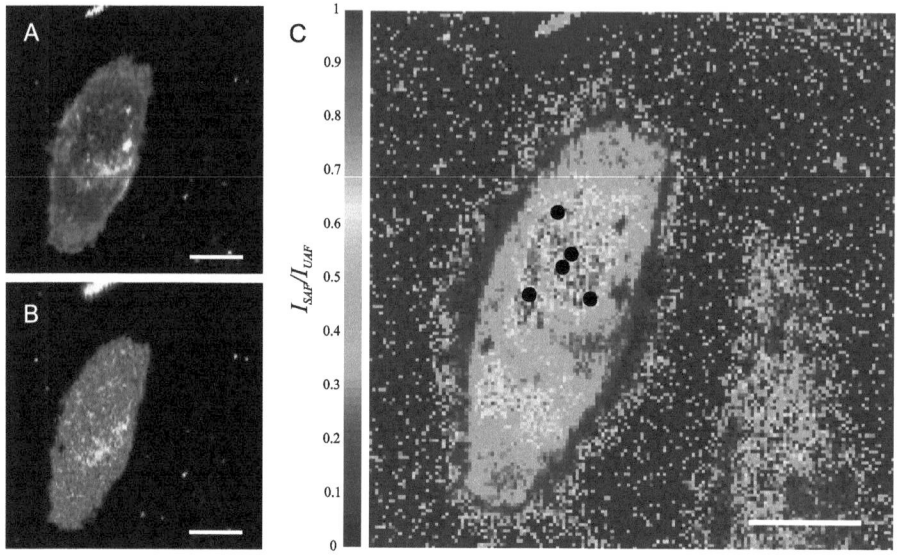

Figure 5.29: Fluorescence intensity scan of HeLa cells stained with Cellmask^{TM}. Images obtained from UAF (A) and SAF (B) in parallel. (C) I_{SAF}/I_{UAF} image after background correction with positions of FCS measurements indicated by the *black circles*. The sample was scanned with a pixel-size of 312.5 nm and 1.3 ms integration time using an excitation intensity of 1 µW. Scale bars=10 µm.

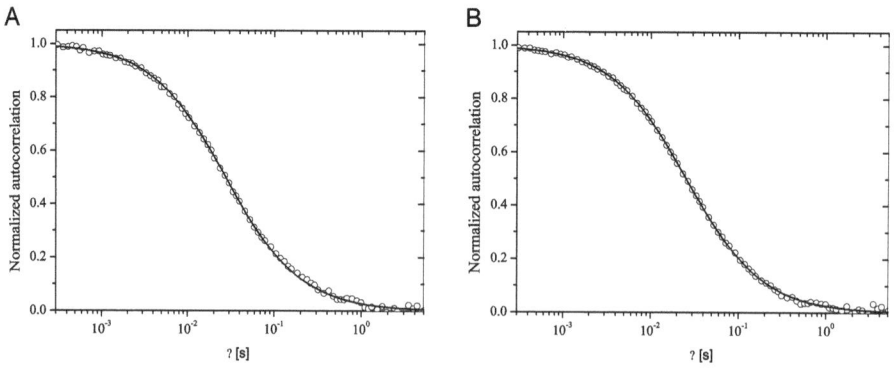

Figure 5.30: Diffusion of Cellmask^{TM} in the plasma membrane of HeLa cells. (*A and B*) Autocorrelation function for SAF and UAF (*open circles*), respectively, and their fits (*solid lines*)

or at least independent from the support.[114] A further aspect to consider is also the location of the lipid dye in the bilayer. It is known that some labelled lipids preferentially locate in one of the leaflets in SLBs.[115] Others have observed a sticking interaction of lipid dyes with the support in SLBs.[116]

5.3 Conclusions and Outlook

In conclusion, with 3D-SAF microscopy a simple method for surface-selective 3D imaging with sub-diffraction-limited resolution has been introduced. As opposed to common near-field optical methods it obviates the need for a probe tip and imaging is not limited to surface features only. A resolution of down to even 1 nm was obtaind with nanoscopic beads. The development of an advanced objective with improved collection efficiency has allowed for extending the method to the axial localization with a resolution of 20 nm for single molecules. The high temporal resolution of the method allowed z-localization with an accuracy of a few tens of nanometers for fast diffusing nanobeads. It is necessary to stress that the method is not restricted to the use of parabolic collectors. The presented concept of simultaneous detection of SAF and UAF is generic and very versatile. The separation of SAF and UAF has already been shown using conventional microscope objectives of sufficiently high NA (>1.45).[117–119] The separation is performed at the back focal plane, which is the angular distribution of intensity. However, it is more difficult to ensure precise separation of SAF and UAF and high angle abberations can pose a serious technical limitation. The current restriction of 3D-SAFM is the diffraction-limited resolution in the focal plane. Several apporaches exist to overcome this problem. One possible means of additionally obtaining lateral super-resolution is to combine 3D-SAFM with STED microscopy (see page 10) due to the conventional excitation optics. Albeit still being diffraction-limited, two-photon excitation could also be employed to enhance the lateral resolution. Another approach for sub-diffraction resolution in the xy-plane is high-precision localization of single fluorophores as in STORM/PALM (refer to page 10). Single-molecule localization microscopy can additionally circumvent the problem that axially overlapping molecules cannot be resolved by 3D-SAFM. However, single-molecule localization in the xy-plane would require wide-field illumination and imaging of the UAF portion with a single-molecule sensitive CCD camera. For a relatively small field of view ($\sim$2 μm in diameter) SAF can still be quantitatively collected with the 180 μm diameter photo-sensitive area of a single-photon avalanche diode (SPAD) and used to determine the z-position. In theory, large z-localization errors can be caused in 3D-SAFM by an anisotropic fluorophore orientation which can also significantly affect the positional accuracy of STORM/PALM-based methods.[120] The experiments on single-molecules have shown that this does not necessarily pose a general obstacle to the method and also in view that approaches exist to determine fluorophore orientation.[121]

The improved collection efficiency and higher signal-to-background brought by the *2-Theta* system enabled the use of parallel detection of SAF and UAF for correlation spectroscopy. A thorough description of the SAF and UAF detection volumes has been presented. The two detection volumes, in particular the volume for SAF, are well defined. The measurement of UAF- and SAF-CS provides access to the bulk concentrations and surface-near concentrations at the same time. This is in particular of advantage for measuring binding equilibria at surfaces and membranes where bulk concentrations are not known. For instance, the measurement of the interaction of proteins with membranes or membrane proteins by FCS often requires the use of fluorescent fusion proteins whose expression levels are not precisely known.[122–124]

A powerful extension of FCS has been introduced for recognizing autocorrelation curves distorted by influences of membrane geometry or vertical membrane movements which would go undetected in its standard realization. It is straightforward to implement techniques for minimizing optical artifacts common to FCS[125] by introducing an internal calibration, as in 2-focus FCS[116] and scanning FCS.[126–128] The method can also be used to support the investigation of complex and in particular anomalous transport by the variable beam-waist approach.[129] Due to the standard excitation optics, also a combination with STED[130] is feasible to further increase the spatial resolution for the study of nanoscale dynamics.[131] The presented concept of simultaneous detection of SAF and UAF could be performed with conventional microscope objectives of sufficiently large NA. This would allow resolving z-axis components of membrane diffusion in imaging-based methods such as SPT.

6 Supercritical Angle Fluorescence Immunoassay Platform*

This chapter will focus on the development SAF immunoassays for the rapid and sensitive quantification of bioanalytes. The assay system comprises single-use test tubes and a fluorescence reader. The mass producible polymer tubes contain an optical configuration for the collection of SAF. A detailed description of the tube and the fluorescence reader can be found in section 4.2. The assay is performed in a sandwich format where fluorescently labelled detection antibodies accumulating at the transparent polymer interface upon formation of sandwich complexes emit SAF [Fig. 6.1A]. A parabolic collector converts SAF into conveniently detectable parallel rays [Fig.6.1B]. With SAF, binding-kinetics are monitored without interference of the fluorescence from the excess detection antibody in solution. Assays are developed for the quantification of three popular analytes, namely interferon-γ, interleukin-2, and parathyroid hormone.

6.1 Tube surface chemistry

The antibody immobilization on the Zeonex® polymer tube substrate was done as described in Ref.[133] and is shown schematically in Fig. 6.2 (refer to page 101 for details of the coating procedure). An important issue to be considered when developing a fluorescence-based biosensor is the contribution to the detected signal caused by the autofluorescence of the material. If the autofluorescence is high or even unstable it can prevent a meaningful readout. With regard to autofluorescence, optical polymers are inferior to high-grade optical glasses and Zeonex®, although superior than other polymers,[134] exhibits some intrinsic autofluorescence. A further contribution to the autofluorescence observed for the developed assay system was the coating of the tubes. A

---

*The results of this chapter were partially published in:
T. Ruckstuhl, C.M. Winterflood, S. Seeger. *Anal. Chem.*, **83**(6), 2345 (2011)[71]
C.M. Winterflood, T. Ruckstuhl, S. Seeger, *Biosensors*, **3**(1), 108 (2013)[132]

CHAPTER 6. SUPERCRITICAL ANGLE FLUORESCENCE IMMUNOASSAY PLATFORM

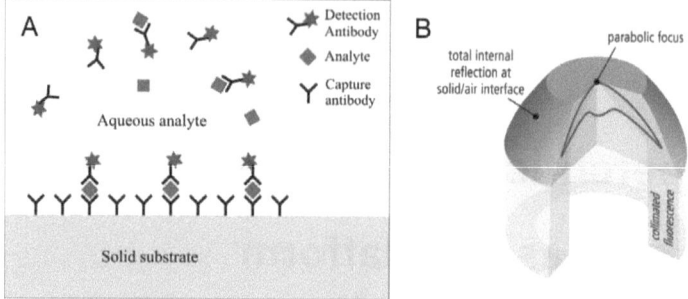

Figure 6.1: (A) Principle of the fluorescence-linked immunosorbent sandwich assay. (B) The SAF collection scheme using a parabolic collector, where the *red line* delineates the angular distribution of surface-bound fluorescence emission.

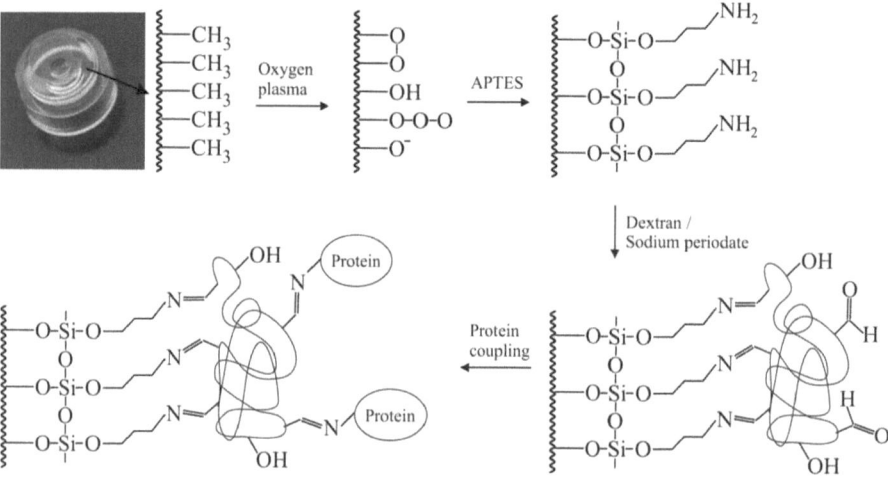

Figure 6.2: Surface chemistry on Zeonex®. The cyclo-olefin polymer surface is oxidized in oxygen plasma and silanized with 3-aminopropyl triethoxysilane (APTES). The resulting amino-terminated surface is subsequently aldehyde functionalized with dextran. The protein (here streptavidin) is imine-coupled to the dextran-matrix via tertiary amines found in N-termini and lysines. A biotin-labelled capture antibody is immobilized via the standard biotin-streptavidin interaction.

quick and simple strategy to minimize the background contribution from the substrate was to photobleach the autofluorescence by light exposure of the polymer before the measurement. This was done by illuminating the coated tubes from below with a 630 nm LED for 1 h at 4°C. The light was focused by the aspheric and parabolic surface producing an irradiance in the order of 100 W/cm^2 at the detection region of the tube [Fig. 6.3]. In order identify sources of autofluorescence introduced

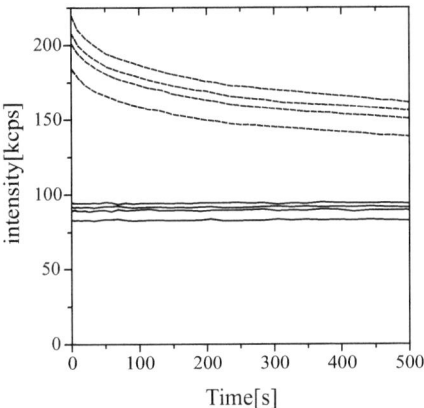

Figure 6.3: Background intensity of eight tubes measured on the prototyped fluorescence reader with (*solid lines*) and without (*dashed lines*) previous exposure of the tubes to LED light for 24 h. Figure from Ref.[71] © 2011 by the Americal Chemical Society.

during the coating procedure and thereby possibly eliminate the necessary photobleaching step before the experiment, the autofluorescence of the tubes was analyzed after each of the coating steps [Fig. 6.4]. The largest contribution of autofluorescence could be ascribed to the silanzation with APTES. Impurities of the employed ethanol and APTES, with them being of analytical grade, could be largely ruled out. A possible source could be contaminants from the used vessels. This might be reflected in the large variation of the autofluorescence after silanization (the tubes analyzed in Fig. 6.4 were not all from the same coating-batch). Due to the high reactivity of the surface after the treatment with oxygen plasma, this step is very critical and should in future be performed in extremely clean vessels, i.e. piranha cleaned glass containers.

CHAPTER 6. SUPERCRITICAL ANGLE FLUORESCENCE IMMUNOASSAY PLATFORM

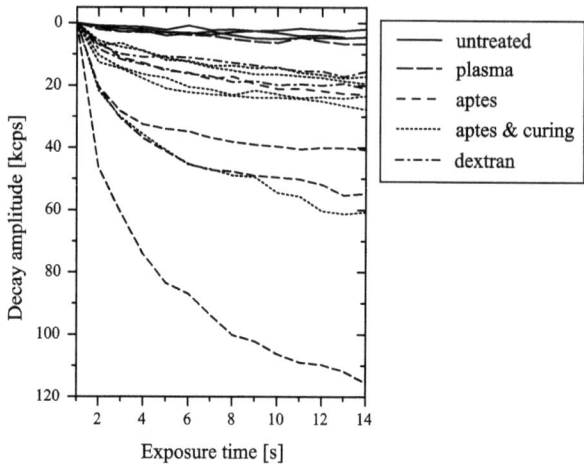

Figure 6.4: Photobleaching decays of the tubes after each coating step over an exposure time of 14 s using 1 mW excitation intensity. Before the corresponding coating step, the tubes were bleached extensively; as such they were "blanked". The autofluorescence was monitored after the treatment with plasma, silanization, curing, and the reaction with dextran.

6.2 Assays

A protocol was established that limits the maximum time requirement of the assay to about 13 min while minimizing the number of working steps. First, forty-five microliters of the analyte dilutions in buffer were pipetted into the tube coated with capture antibodies. Immediately after, 5 µl detection antibody solution was added and the solution was mixed briefly with a pipet. The tube was then inserted into the reader instrument and the measurement was started. The measurement was controlled using a customized software from a laptop and required no further manual intervention. Due to the high surface selectivity of the SAF method, washing steps could be omitted in the assay protocol. During the first 700 s, the formation of the sandwich complexes at the surface was monitored by measuring the increase of the SAF intensity with time. An excitation intensity of 1 µW was used, producing a low irradiance of 0.05 W/cm^2 within the illuminated spot at the surface. Photobleaching of surface bound fluorophores was kept well below 1% by measuring the SAF intensity only intermittently with a laser exposure time of 1 s. The binding followed rather complex kinetics as the sandwich formation at the surface could proceed through two pathways, with analyte molecules binding either to the detection antibody in solution first or to the capture antibody on the surface. The SAF intensity approaches a saturation level for analyte concentrations of around 2/3 of the detection antibody concentration as a result of the depletion of the detection antibody. Rather low concentrations of detection antibody were chosen (1.5 nM-15 nM) to optimize the conditions for the detection of very low analyte concentrations. Using higher concentrations of the detection antibodies, the upper detection-limit can be extended by almost 2 orders of magnitude until depletion of the free capture antibodies on the substrate. For the measurement of very low analyte concentrations, it was necessary to use a higher excitation intensity to increase the detectable fluorescence signal. The excitation intensity was boosted by a factor of thousand from 1 µW to 1 mW. This led inevitably to an increased photobleaching of the surface-bound detection antibody. The amplitude of the intensity decay obtained after extensive photobleaching of the surface-bound fluorophores serves as a precise measure for the amount of fluorescent analyte present at a surface.[24,74] The background contribution of the measurement substrate can vary from one tube to the next. For sensitive measurements, it is usually necessary to perform blank measurements and subtract the background intensity from the fluorescence signal. For routine use, however, such procedure is cumbersome increasing the amount of work for the user or the level of technical complexity for automated assay systems. This issue was circumvented by the photobleaching method where the intensity decays are independent of the photostable background of the substrate. This method was employed by increasing the excitation intensity to 1 mW, 700 s after starting the measurement, allowing enough time for sufficient analyte molecules to bind to the surface. Over 95% of the surface-bound fluorophores were bleached after a total exposure time of just 11 s.

Interferon-γ assay

Interferon-γ (IFN-γ) is a small $\sim$16 kDa homodimeric cell-signalling protein that is critical for innate and adaptive immunity against viral and intracellular bacterial infections and for tumor control. Aberrant IFN-γ expression is associated with a number of infectuous, autoinflammatory, and autoimmune diseases and has shown therapeutic activity for the treatment of rheumatic diseases.[135,136] The sensitive and accurate determination of IFN-γ concentrations is of great importance in immunological research and medical diagnostics. Sandwich assays were performed for the quantification of recombinant mouse IFN-γ (Invitrogen) in buffer. Several combinations of capture and detection antibodies were tested as there is some reluctance as to the disclosure of suitable antibody pairs due to commercial interests. The following combinations of rat anti-mouse monoclonal biotinylated capture and Cy5-conjugated detection antibody clones were tested and gave unsatisfactory results: Biotin-XMG1.2/Cy5-R4-6A2, biotin-R4-6A2/Cy5-XMG1.2, biotin-AN-18/Cy5-R4-6A2.

Interestingly, the assay was successfull when capture and detection clones AN-18 and R4-6A2 were swapped to give biotin-R4-6A2/Cy5-AN-18. Figures 6.5 and 6.6 show the results obtained for the low and high sensitivy read-out mode using a final concentration of 10 nM detection antibody with a dye to protein ratio of 1.7.

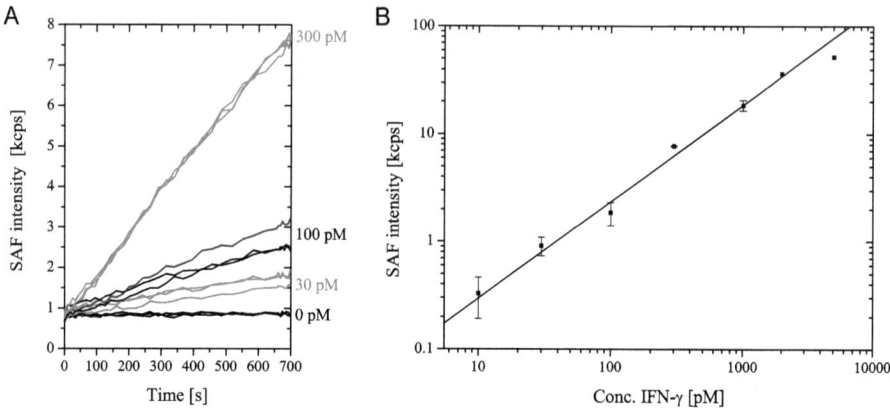

Figure 6.5: (A) Real-time measurements for different IFN-γ concentrations. (B) Increase of the SAF intensity after 700 s plotted against IFN-γ concentration. A straight line through the origin was fitted to the data for IFN-γ concentrations up to 2 nM. At IFN-γ concentrations higher than 2 nM the linear relationship was lost due to depletion of the free detection antibodies in solution.

Smooth binding curves could be obtained already for picomolar analyte concentrations thanks to the excellent signal-to-background ratio of the system. The limit of detection (LOD, see page 101 for explanation) was calculated from the sensitive read-out mode from the intersection of the linear fit with the 3-σ value of the zero concentration measurements to be 1.9 pM (30.4 pg/mL). According to the supplier of the antibodies (eBioscience, San Diego, CA, USA), their recombinant standard range for the equivalent ELISA is 15-2000 pg/ml with an assay time of 4 h. In comparison, the SAF-assay provided a linear response over a concentration range of 30.4 pg/ml to 32 ng/ml in 13 min involving only two consecutive liquid additions. The mean coefficient of variation (CV) over all measurements was 14.8%.

Interleukin-2 assay

Interleukin-2 (IL-2) is small ~17 kDa cell-signalling protein produced by T lymphocytes which is involved in the body's natural response to microbial infection and in discriminating between foreign

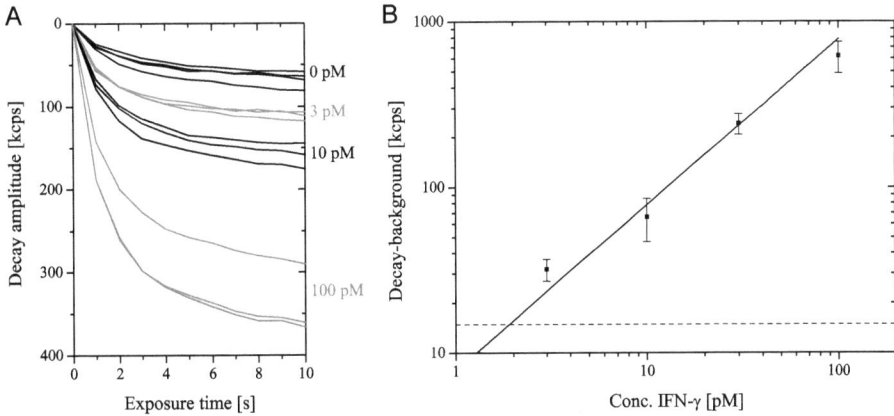

Figure 6.6: IFN-γ sensitive read-out after 700 s with 1 mW excitation. (A) Photobleaching decays of the SAF intensity. (B) Plot of the photobleaching amplitudes after 11 s minus background (zero concentration decay) against IFN-γ concentration. A straight line through the origin was fitted through the datapoints. The *horizontal line* represents the 3-σ value of the blank measurement.

and self. In medical diagnostics, IL-2 serum levels provide insight into a myriad of pathological situations. Sandwich assays were performed for the quantification of recombinant mouse IL-2 (Invitrogen) in buffer. The assay was performed with biotinylated rat anti-mouse IL-2 monoclonal antibody clone JES6-5H4 (eBioscience) as capture antibody and Cy5-labelled rat anti-mouse IL-2 monoclonal antibody clone JES6-1A12 (eBioscience) as detection antibody. Figures 6.7 and 6.8 show the results obtained using a final concentration of 1.5 nM detection antibody with a dye to protein ratio of 2.6.

The LOD was 0.27 pM (4.5 pg/ml). The SAF-assay had a mean CV of 12.4% and a linear response over a concentration range of 4.5 pg/ml to 10 ng/ml. In comparison, the supplier of the employed antibodies claims a recombinant standard range for the equivalent ELISA of 4-500 pg/ml with an assay time of 4 h.

Parathyroid hormone assay

Parathyroid hormone (PTH) is a small ~9 kDa 84 amino acid (aa) protein hormone secreted by parathyroid glands and is involved in the regulation of calcium levels in the blood. This specific molecule was chosen due to the clinical importance of its rapid and sensitive quantification.

CHAPTER 6. SUPERCRITICAL ANGLE FLUORESCENCE IMMUNOASSAY PLATFORM

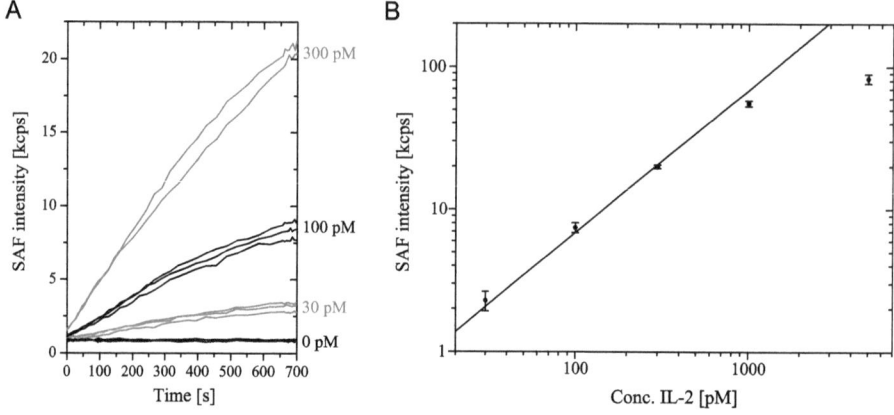

Figure 6.7: (A) Real time measurements for different IL-2 concentrations. (B) Increase of the SAF intensity after 700 s plotted against IL-2 concentration. A straight line through the origin was fitted to the data for IL-2 concentrations up to 1 nM. The linear relationship was lost at concentrations higher than 1 nM due to depletion of detection antibody.

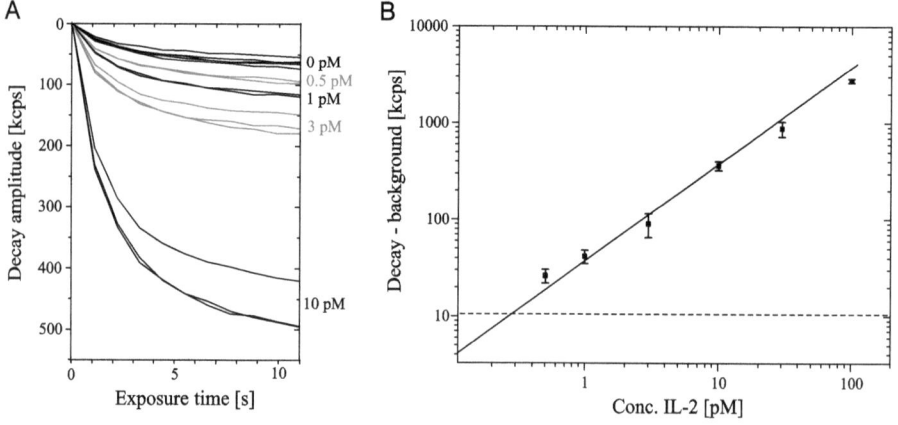

Figure 6.8: IL-2 sensitive read-out after 700 s with 1 mW excitation. (A) Photobleaching decays of the SAF intensity. (B) Plot of the photobleaching amplitudes after 11 s minus background (zero concentration decay) against IL-2 concentration. A straight line through the origin was fitted through the datapoints. The *horizontal line* represents the 3-σ value of the blank measurement.

Parathyroid adenoma leads to elevated levels of PTH in the blood. During parathyroid-ectomy PTH is measured intraoperatively to confirm the complete removal of the adenoma.[137] Intact PTH (iPTH) serves as a good marker, as it has a half-life of only 2–5 min being rapidly hydrolized into various fragments once released into the blood stream. In order to measure exclusively iPTH, two antibodies are used; one is directed only towards the N-terminal region and the other towards the C-terminal region. In a first trial experiment iPTH measurements were performed in buffer. Recombinant human iPTH (ProSpecBio, East Brunswick, NJ, USA) was quantified using a biotinylated goat anti-human PTH C-terminal specific (aa 53–84) polyclonal capture antibody (Lifespan Biosciences, Seattle, WA, USA) and a Cy5-labelled monoclonal N-terminal specific (aa 1–34) mouse anti-human PTH clone BGN/1F8 (Biotrend, Köln, Germany) detection antibody labelled with a dye to protein ratio of 1.0. The results of the assay are shown in Fig. 6.9.

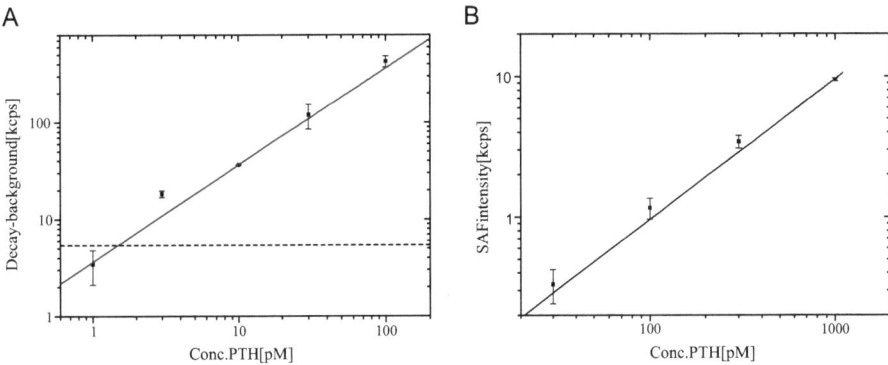

Figure 6.9: Calibration curves for PTH obtained in buffer for (A) the sensitive read-out mode from the photobleaching decay amplitude, where the *horizontal line* represents the 3-σ value of the blank measurement, and (B) from the fluorescence intensity after 700 s.

The assay achieved a limit of detection of 1.5 pM (15 pg/ml) with a mean CV of 14.7%. The concentration range of healthy individuals is 10–60 pg/ml. The assay sensitivity in buffer was high enough to distinguish between normal and elevated levels and it was proceeded to measure hormone concentrations in patient sera with normal and elevated PTH levels provided by the Universitätsspital Zürich. Unfortunately, there was no correlation between our measurements and those performed with the same samples on a Cobas Elecsys E170 system (Hoffmann-La Roche, Basel, Switzerland) at the Universitätsspital. However, the patient sera could be supplemented with recombinant PTH down to few tens of picomolar and the SAF measurements gave reliable results. It remained unclear to what this could be ascribed to. Possible issues could have been cross-reactivity with truncated PTH or non-specific interaction with serum components. Variations in autofluorescence of sera from human to human could be ruled out as a cause.

6.3 Enhancement of the Assay Sensitivity

A straightforward improvement of the assay sensitivty can be achieved by extending the time allowed for analyte molecules to bind to the surface until read-out. The fluorescence signal scales in good approximation linearly with the measurement time as the binding capacity of the surface is by far not exhaused for low analyte concentrations. A further strategy would be to increase the mass transport towards the surface as the formation of sandwich complexes is essentially limited by diffusion. During the assay, the high surface density of capture antibodies leads to a rapid depletion of the analyte in the surface-near region thereby reducing the local concentration and rate of the increase in fluorescence signal. The sensitivity could be increased by a factor of ∼2.5 if the tubes were agitated in a circular motion about their axis as shown in Fig. 6.10.

A very noteworthy and distressing fact is that 99.98% of the receptors on the surface lie outside

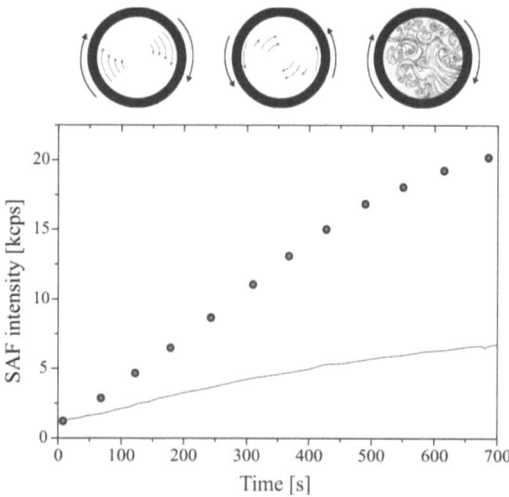

Figure 6.10: Comparison of the time-dependent increase in fluorescence intensity between an assay performed with circular agitation (*circles*) and under static conditions (*line*) with 100 pM IL-2.

the detection region. For simplicity of the immobilization procedure, the coating was applied over the entire interface between the substrate and solution with 4 mm diameter. A confinement of the receptors to the detection region reduces the effect of analyte depletion near the surface as can be shown by a simple simulation shown in section 6.3.

Simulations on the effect of capture spot size

A monte-carlo approach was used to simulate the effect of confining the capture antibodies to the detection spot. The liquid container of the tube was approximated by a cylinder with a base of 4 mm diamter and 2 mm height (50 µl). The limiting step of the assay is the binding of the detection antibody to the surface, either in form of a complex with the analyte or alone. The system was thus reduced to the diffusion of a single entity with the diffusion coefficient D=40 $\mu m^2 s^{-1}$ of an IgG antibody, which is in good approximation also the diffusion coefficient of the analyte-detection antibody complex. Diffusion was simulated as a Gaussian random-walk in three dimensions with continuous position variables and reflective boundary conditions. N molecules were initialized at random in the cylinder and moved in each coordinate by a random value drawn from a Gaussian distribution with the standard deviation $\sigma=\sqrt{2D\Delta t}$ for every timestep Δt=10 ms. If the particle was found at $z \leq 0$ and at a radial position $r \leq r_d$, where r_d is the radius of the detection spot, the molecule was bound and the SAF intensity was incremented by one. If found at $z \leq 0$ and at a radial position $r_d \leq r \leq r_b$ it was simply bound. The simulations in Fig. 6.11A show that within less than 30 s the surface-near concentration is reduced to less than ~10% with respect to the bulk in the case where the capture region extends over the entire surface. In the case of a capture spot of only 100 µm diameter, the concentration over the detection area is reduced only to ~45% [Fig. 6.11B]. Figure 6.11C shows the time-dependence of the number of binding molecules. Without confinement of the capture area the binding curve begins to flatten off after already 300 s due to depletion of analyte in solution, while for the confined capture-area there is no decrease in the binding-rate. The theoretically achieved sensitivity is over twofold for a read-out after 600 s for a confined capture area.

Towards reducing the capture-spot size by photolithography

A very elegant approach to restrict the capture antibody to only the detection region is to use the focusing lens of the tube for photolithography thereby confining the immoblization to precisely the excitation spot. Nitroveratryloxycarbonyl (NVOC) is an established photolabile protecting group. It can be used to deprotect primary amines by irradiation with 350 nm UV-light – a wavelength at which the Zeonex® material is still highly transmissive. The key idea is to use an NVOC derivative

CHAPTER 6. SUPERCRITICAL ANGLE FLUORESCENCE IMMUNOASSAY PLATFORM

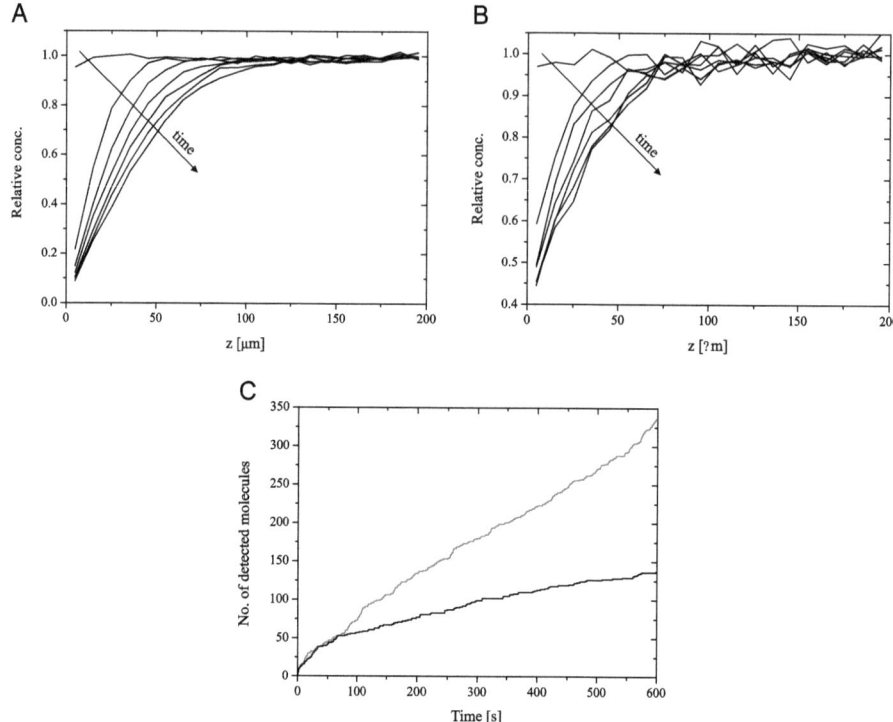

Figure 6.11: Time dependence of the surface near concentration in steps of 5 s, starting from an initial concentration of 1 pM ($\sim 3 \times 10^7$ molecules) for capture antibodies covering the entire surface (A) and covering only a 100 µm diameter disk over the 50 µm diameter detection area (B). In (A) the concentration was averaged along the entire horizontal plane, in (B) only above the 50 µm diameter detection area for 40 trajectories for better statistics. (C) Time-dependence of the number of detected molecules for capture antibodies covering the entire surface (*black curve*) and for a 100 µm diameter capture spot (*gray curve*).

of APTES, which is used in the first coating step of the tubes, to introduce amine functionality (-NH_2). The substrate coupled NVOC-APTES is then UV-irradiated to produce the desired free amines [Fig. 6.12]. The coating would then proceed as usual. The use of NVOC-APTES in a similar context has been reported in literature.[138,139] NVOC-APTES was not commercially available and was therefore synthesized as described on page 103. Figure 6.13A shows a UV/Vis-spectrum of the synthesized NVOC-APTES before and after irradiation with UV-light. For this, a solution of

CHAPTER 6. SUPERCRITICAL ANGLE FLUORESCENCE IMMUNOASSAY PLATFORM

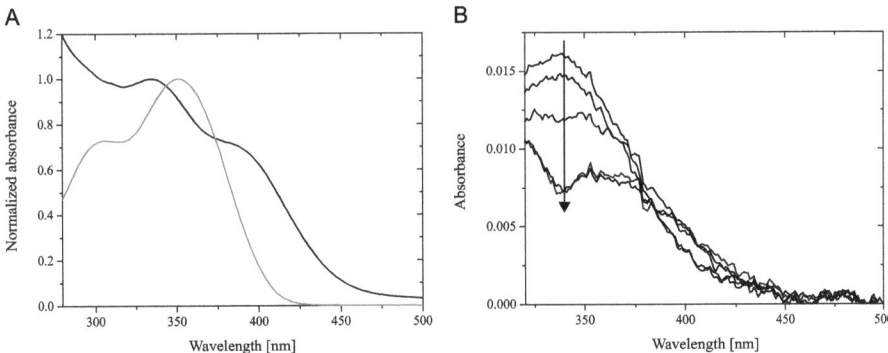

Figure 6.12: Principle of photolithography where NVOC-APTES is deprotected only in the detection spot by irradiation with UV-light using the asphere of the tube. The deprotection reaction of the amine is a light-induced intramolecular oxidation of the benzylic carbon-hydrogen bond in *ortho* position to the nitro group, which leads to an aromatic aldehyde.

NVOC-APTES in DMF in a quartz cuvette was exposed for different times with a UV mercury lamp (irradiance: 35 mW/cm^2). For the coupling of NVOC-APTES to Zeonex®, trial experiments were performed on planar slabs of the polymer with 1.5 mm thickness (see page 103). The success of the coupling reaction was confirmed by UV/Vis spectrophotometry [Fig. 6.13B]. The obtained spectra were very similar to spectra found in literature.[140,141] To assess the amine functionality after

Figure 6.13: (A) UV-spectra of NVOC-APTES in DMF solvent before (*gray*) and after exposure to UV-light (*black*). (B) UV-spetra of NVOC-APTES coupled to Zeonex® after different irradiance times with a 350 nm UV-LED.

photolysis, a 1 mm diameter circle was exposed to a 1 mW 350 nm UV-LED using a mask and subsequently incubated with 5 µM of amine-reactive Cy5-NHS in 0.1 M sodium carbonate, pH 8.3 for 1 h. Figure 6.14 shows the resulting fluorescence image obtained using a wide-field fluorescence microscope. The successful deprotection of NVOC-APTES on Zeonex® could be demonstrated. However,

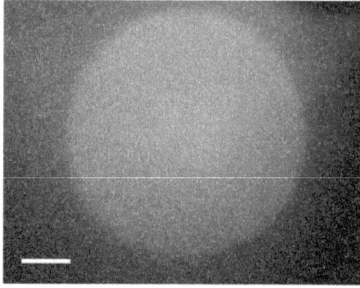

Figure 6.14: Fluorescence image of site-directed coupling of Cy5 on a 1 mm diameter circular area on Zeonex® substrate using photo-deprotection of NVOC-APTES. Scale bar=200 μm.

preliminary experiments for the coupling of dextran to the deprotected amine, as described on page 101, failed and needs to be further investigated.

6.4 Conclusions and Outlook

The SAF approach combines high collection efficiency with high surface selectivity making it ideal for the read-out of solid-phase immunoassays. The presented immunoassays show a sensitivity and linear concentration range comparable to the ELISA with a turnaround time of less than 15 min and only involving two consecutive liquid additions. It can very well be reduced to one addition by depositing lyophilized detection antibodies inside the tube at the production process. The method doesn't rely on the common indirect detection with a secondary antibody and enzymatic amplification. The assay therefore requires a minimal amount of material and human intervention, which is a major advantage especially in terms of consistency and reproducibility of results. The main contribution to the CVs of the measured assays could be traced back variations in the receptor density from one tube to another resulting from the laborious manual immobilization procedure. Improvement in terms of speed can be expected by introducing a mass transport towards the surface and confining the capture spot to the detection area. A first step towards central immobilization by photolithography using the exctiation optics has been made. In spite of all, one of the most crucial factors for the performance of an assay still remains the affinity and specificity of the employed antibodies. The proprietary technology for the capture of SAF in single use receptacles is suitable for integration into well plates.[142] The optical configuration tolerates a substantial lateral mismatch in the order of several hundred micrometers between its symmetry axis and the excitation beam, and rapid sequential readout of numerous wells can be done using inexpensive actuators [Fig. 6.15]. For comparison, with the focusing element included in the reader instead of the receptacle, the lateral tolerance of the alignment is only a few tens of micrometers.[16]

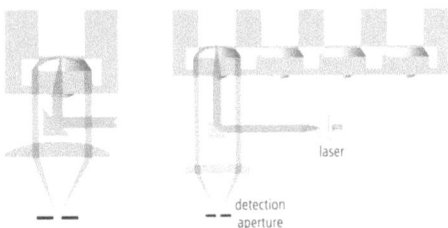

Figure 6.15: A suggestion for parallelizing SAF-assays. The integration of the excitation optics makes it extremely tolerant towards the lateral alignment of the laser beam allowing for rapid sequential read-out of multi-well plates.

7 Material and Methods

3D-SAFM: Fibroblast culturing and immunofluorescence staining

NIH 3T3 mouse embryonic fibroblast cells (a courtesy of the Institute of Molecular Biology, University of Zürich) were grown in cell culture dishes in an incubator (Sheldon TC2322, Cornelius, OR, USA) under 5% CO_2 and at 37° to 50% confluency in full DMEM medium supplemented with 15 mg/l Phenol-red, 2 mM L-glutamine, 4.5 g/l D-glucose and 10% fetal calf serum (Invitrogen). The cells were briefly rinsed with prewarmed Tris-buffered saline (20 mM Tris-HCl, pH 7.4) and incubated for 3 min in 0.5 g/l trypsin, 0.2 g/l EDTA (Sigma-Aldrich) at 37°C to detach the cells from the culture dish. A fourfold volume of full DMEM media prewarmed to 37°C was then added and the cell suspension was transferred to the microscope coverslip, which had been incubated for 30 min in 50 µg/ml fibronectin (Invitrogen). The cells were left to adhere for 2-3 h, and then incubated for 30 min at 37°C in full DMEM with 10 µM nocodazole (Sigma-Aldrich), 30 min at 4°C and further 30 min at 37°C in full DMEM. The coverslips with cells were rinsed with PBS (Invitrogen) and the cells were stained by indirect immunofluorescence according to a modified protocol from Ref.[143] In brief, the cells were extracted for 3 min in microtubule-stabilizing lysis-buffer (60 mM PIPES, 25 mM HEPES, 10 mM EGTA , 2 mM $MgCl_2$, pH 6.9 supplemented with 0.5% (v/v) Triton-X-100 (Fluka) and 10 µg/ml paclitaxel (Sigma-Aldrich) prewarmed to 37°C. For fixation the cells were incubated in 1% (v/v) glutaraldehyde in PBS for 30 min and incubated in 1% (w/v) sodium-borohydride in PBS for 3 min. The cells were blocked for 1 h in 3% (w/v) IgG-free bovine serum albumin (Dianova, Hamburg, Germany) in PBS. The cells were incubated for 1 h in 1 µg/ml bovine anti–tubulin mouse monoclonal antibody, clone DM1A (Dianova) in PBS followed by incubation for 1 h in 1 µg/ml Cy5-conjugated goat-anti-mouse IgG (F_c-fragment specific) polyclonal antibody (Dianova) in PBS.

3DSAFM: Microsphere sample preparation

Non-porous silica microspheres with a mean diameter of 4.74 µm (10-15% coefficient of variation) (Bangs Laboratories, Fishers, IN, USA) were fluorescently labeled by immersion in a saturated

CHAPTER 7. MATERIALS AND METHODS

ethanolic solution of DiIC$_{18}$(5) (Invitrogen) followed by multiple steps of 2 min centrifugation at 1000 g (Heraus Labofuge 400R, Hanau, Germany), removal of the supernatant, and resuspension in ddH$_2$O. The last step was done in an index-matching solution of glycerol/water. The RI of the microspheres was determined by the immersion method using a transmitted-light microscope (Zeiss, Model 464002-9901, Oberkochen, Germany) using a 16× magnification. The principle of the method works as follows:

1. The microscope was placed in the 23°-thermostated room where also the measurement with the fluorescent microscope was performed to avoid any influences of the temperature on the RI.

2. 0.5 µl the of the aqueous bead solution as delivered was dissolved in 1.5 ml of a glycrol/water solution with an RI close to the suspected RI of the beads (RI 1.43 claimed by BangsLabs).

3. A ~25 µl drop of the bead-solution was placed on a clean microscope slide.

4. If while adjusting the focus in a direction that decreased the distance between the microscope stage and the objective lens, a dark ring appeared on the circumference of the beads and light concentrated in the center, the solution used had an RI higher than that of the beads. Adjustment of the focus in the opposite direction showed the beads blurring, with no ring and a bright center spot appearing. If instead while adjusting the focus in a direction that increased the distance between the microscope stage and the objective lens, a dark ring appeared on the circumference of the beads and light concentrated in the center, the liquid used had an RI lower than that of the beads. Adjustment of the focus in the opposite direction shows the beads blurring, with no ring and a bright center spot appearing.

5. This evaluation and selection of glycerol concentrations with different RIs was continued until the beads were almost invisible when perfectly in focus. When perfectly matched, they had a blurry outline when defocused in either direction.

By doing so, a solution of 66.25% (v/v) of glycerol/water was found to match the RI of the beads. The RI of the solution was then determined with an Abbe refractometer (Atago 1T, Tokyo, Japan) thermostated at 23° to be 1.427 ± 0.002. The wavelength used by the refractometer is the sodium D line at 589 nm. The dispersivity of the silica beads and the solution was not taken into account, as the measurement was performed using 633 nm excitation. The coverslip for the fluorescence microscopy measurement was cleaned by sonication for 20 min in 100% ethanol, rinsed with ddH$_2$O, piranha-cleaned (50% H$_2$SO$_4$/15% H$_2$O$_2$/35% H$_2$O) for 20 min, and rinsed with ddH$_2$O and dried under nitrogen flow.

3DSAFM: Nanobeads sample preparation

Carboxylate-modified polystyrene beads with a mean diameter of 36 nm (20% coefficient of variation) and RI of 1.59 were purchased from Invitrogen. The coverslip was cleaned by sonification for 20 min in 100% ethanol, rinsed with ddH$_2$O, piranha-cleaned (50% H$_2$SO$_4$/15% H$_2$O$_2$/35% H$_2$O) for 20 min, and rinsed with ddH$_2$O and dried under nitrogen flow. The beads were dispersed by sonificating them for 20 min before use. For adsorption to the glass coverslip, the beads were diluted into 50 mM sodium-citrate, pH 3.0. In their protonated, neutral form the beads have a high tendency of adsorbing to the coverslip. For the preparation of beads suspended in agarose the beads were diluted into a solution of 1% (w/v) agarose (Inno-Train, Kronberg, Germany) heated above gelling temperature in a microwave. The suspension was immediately spin coated with an in-house made device to a thin film of less than 1 µm thickness. The RI of the gelled 1% solution of agarose was measured with an Abbe refractometer (Atago 1T) to be 1.333 ± 0.001. For imaging, the agarose film was overlaid with ddH$_2$O water.

FCS: Detection volume characterization – sample preparation

The coverslips were first rinsed with ethanol and 1% (v/v) Deconex®, glued to the sample holder and rinsed again with ethanol and 1% (v/v) Deconex®. The coverslips were then treated with oxygen plasma using a Femto plasma device (Diener electronic, Ebhausen, Germany). Improper coverslip preparation can quickly lead to adsorption effects as shown in Fig. 7.1. To prepare a very

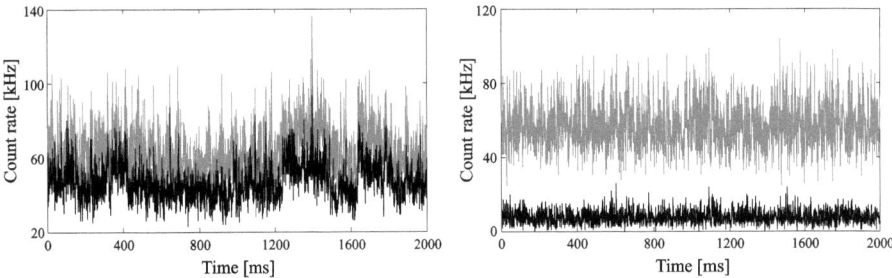

Figure 7.1: SAF (*black*) and UAF (*gray*) intensity tracks with (*left*) and without non-specific adsorption (*right*) to the coverslip glass. The intensity tracks on the left were recorded with a coverslip which was not plasma-treated. The sample was 10 nM Atto655 in 200 mM NaCl. The excitation intensity was 13 µW.

accurate dilution of Atto655 (Sigma-Aldrich), the dye was first diluted as far as possible for it to

CHAPTER 7. MATERIALS AND METHODS

be still measured accurately by UV/Vis spectrophotometry; for a Lambda 900 spectrophotometer (PerkinElmer, Waltham, MA, USA): OD>0.1. The extinction coefficient ϵ_{650}=110'000 M^{-1}cm^{-1} was used according to the supplier. The solution was then further diluted using volumetric flasks while always keeping volumes larger than 100 µl.

FCS: Preparation of supported lipid bilayers

A mixture of 65mol% DOPC and 35mol% DOPS (Avanti Polar Lipids Inc., Alabaster, AL, USA) in chloroform was stirred under vacuum to remove the solvent and left for another hour under vacuum (< 10 mbar). The lipids were resuspended to 1 mg/ml in membrane buffer (100 mM NaCl, 3 mM CaCl$_2$·3H$_2$O, 10 mM Tris, pH 7.5) (Sigma-Aldrich) and extruded with a Mini-Extruder (Avanti Polar Lipids) at least 40 times through a 0.1 µm pore size membrane (Whatman, Maidstone, UK) to produce unilamellar lipid vesicles with a homogeneous size distribution. The SLBs were prepared on a piranha-cleaned (the piranha solution was composed of one part 35% hydrogen peroxide and three parts concentrated sulfuric acid) glass slide with 0.1 mg/ml of lipids in membrane buffer. The membrane was rinsed to remove unfused vesicles. The membrane intercalating fluorophore Cellmask^{TM} (Invitrogen) was added after complete formation of the bilayer and the bilayer was rinsed extensively with membrane buffer before the FCS measurements.

FCS: Preparation of fibroblast and HeLa cells

p53-deficient mouse embryonic fibroblast cells and HeLa cells (a courtesy of the Institute of Molecular Life Sciences, University of Zürich) were grown in cell culture dishes in an incubator (Sheldon TC2322) under 5% CO_2 and at 37° in full Dulbecco's modified eagle medium supplemented with 15 mg/L phenol-red, 2 mM L-glutamine, 4.5 g/L D-glucose and 10% fetal calf serum (Invitrogen). The cells were detached from the culture dish by brief incubation in 0.5 g/L trypsin, 0.2 g/L EDTA (Sigma-Aldrich). A fourfold volume of growth medium was added and the cell suspension was transferred to oxygen-plasma treated microscope coverslips. The cells were grown overnight and the growth medium was replaced by Hank's balanced salt solution (HBSS, Invitrogen) for FCS measurements. HBSS being CO_2 independent for pH regulation allowed for culture work outside of the regulated atmosphere of an incubator. The cells were incubated for at least 15 min with the plasma membrane stain Cellmask^{TM} and rinsed extensively prior to the FCS measurements. The labelling extent of the cells was hard to control, as it was dependent on the cell density. Therefore, a series of dye concentrations in the lower micromolar range was used. The measurements were performed at room temperature.

SAF Immunoassay: Limit of detection

Several quantities are used to describe the performance characteristics of a method for analyte quantification. The most commonly used is the LOD (limit of detection). According to IUPAC, the LOD is defined as the minimum amount of a substance that can be distinguished from a blank measurement within a confidence limit of 1%. For normally distributed measurements, the LOD is three times the standard deviation σ of the blank measurement. For a signal at the LOD, the probability for a false positive (α error) is 1%, while the probability for a false negative (β error) is 50% for a sample with a concentration at the LOD.

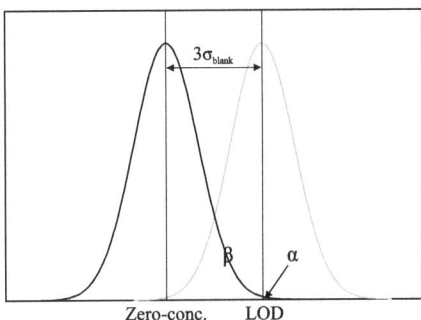

Figure 7.2: Illustration of the limit of detection, where *black* and *gray* are the normal distributions associated with the blank and the limit of detection, respectively.

SAF immunoassay: Protein immobilization procedure

Before assembly of the tubes, the Zeonex® substrates [Fig. 7.3] were activated by oxygen plasma (40 kHz/ 100 W/0.2-1 mbar) using a Femto plasma device (Diener Electronic, Ebhausen, Germany) for 5 min and silanized by immersion in a 3% (v/v) solution of 3-aminopropyl triethoxysilane (Sigma-Aldrich, St. Louis, MO, USA) in ethanol for 2 h. The tubes were rinsed with ethanol and water, dried under nitrogen flow, and left to cure overnight. The silanized Zeonex was functionalized with aldehyde-activated dextran by Shifft's base coupling. For this, it was immersed in a solution of 2% (w/v) dextran T40 (Carl-Roth, Karlsruhe, Germany) and 30 mM sodium periodate (Sigma-Aldrich) for 2 h, rinsed with ddH$_2$O, and further oxidized in 30 mM sodium periodate (Sigma-Aldrich) for 2 h. The Zeonex® substrates were assembled with the O-ring and the upper tube part. Streptavidin (Sigma-Aldrich) was immobilized by Shiff's base coupling by filling the

CHAPTER 7. MATERIALS AND METHODS

tubes with 50 µl of a 1 mg/ml solution of phosphate buffered saline (0.01 M PBS, pH 7.4) and incubating them overnight at 4 °C. The tubes were further incubated with 100 µl of 5 mM glycine/PBS to block unreacted aldehydes. The tubes were washed several times with 0.05% (v/v) Tween 20 (Sigma-Aldrich) in PBS. For later use, the tubes were incubated with 100 µl of Liquid Plate Sealer® (Candor Bioscience, Wangen, Germany) for 1 h at 4 °C. The solution was removed and the tubes were dried under nitrogen flow. After this treatment the tubes could be stored at 4 °C in dry conditions for a longer period. For the presented experiments the storage time was within 2 weeks. For tubes that were stored for 5 months before the immobilization of capture antibodies, the SAF intensity increase measured during the immunoreactions was reduced by about 20%. Prior to the assays, the tubes were incubated for 1 h with 50 µl of biotinylated capture antibody at 30 µg/ml in PBS and washed several times with 0.05% (v/v) Tween 20 in PBS. The tubes were blocked for 1 h with 3% (w/v) bovine serum albumin/0.05% (v/v) Tween 20 in PBS. The assays were performed in 3% (w/v) bovine serum albumin/0.05% (v/v) Tween 20 in PBS.

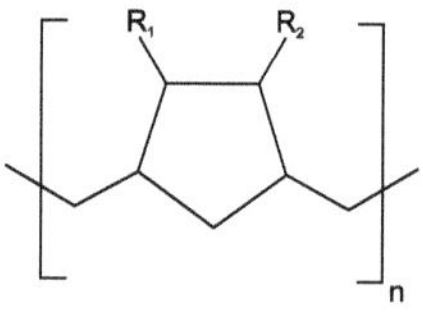

Figure 7.3: Chemical structure of Zeonex®.[144] Zeonex is synthesized by ring-opening metathesis polymerisation of cyclic olefinic monomers and subsequent hydrogenation. Hydrogenation removes carbon-carbon double bonds thereby increasing the thermal stability of the plastic.

SAF Immunoassay: Antibody labelling

The employed antibodies were not all available as fluorophore or biotin conjugates and were labelled by standard N-hydroxy succinimidyl (NHS) ester coupling chemistry [Fig. 7.4]. In brief, the

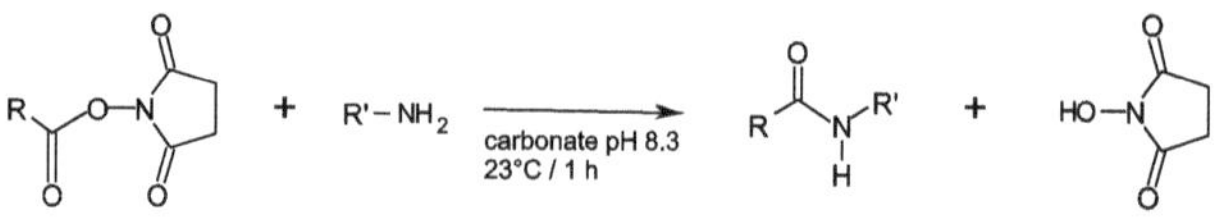

Figure 7.4: Reaction scheme for antibody coupling to the NHS-ester derivative of Cy5, were R and R' represent the fluorophore and the antibody, respectively.

buffer in which the antibodies were delivered was exchanged for labelling buffer (0.1 M sodium carbonate, pH 8.3) using Microcon® 3 kDa molecular weight cut-off centrifugal filters (Millipore, Billerica, MA, USA) spun at 14'000 g in a Heraus Labofuge 400R for 25 min at 4°. Concentration from 500 µl to 35 µl and topping up with new buffer was performed thrice. The concentration of the antibody was then measured with a Lambda 900 spectrophotometer (PerkinElmer) using the absorbance at 280 nm of a typical IgG-antibody (ϵ_{280}=150'000 M^{-1}cm^{-1}). For labelling the concentration was kept around 1 mg/ml. In the case of fluorophore labelling, the antibodies were labelled with a ten-fold molar excess of Cy5-NHS (Invitrogen, Carlsbad, CA, USA) to give a final dye labelling degree of one to three. In the case of biotin labelling, the antibodies were labelled with a three-fold molar excess of biotin-NHS (Invitrogen) to give a final dye labelling degree of around one. Over-labelling with dye or biotin can either cause self-quenching of the fluorophores and/or the loss of the biological activity of the antibodies.[145] The labelling reaction was performed at room temperature for 1 h. Unconjugated dye and biotin was then removed by buffer exchange into 0.1 M PBS using centrifugal filters (5 cycles of buffer exchange). While it was not possible to determine the labelling degree of biotin directly by spectrophotometry, the labelling degree of the fluorophore was determined according to

$$\text{dye/protein} = \frac{A_{650} \cdot \epsilon_{280}}{\epsilon_{650} \cdot (A_{280} - r \cdot A_{650})},$$

where A is the respective absorbance of the dye-antibody conjugate, ϵ_{280} and ϵ_{650} are the extinction coefficient of the antibody and Cy5 (ϵ_{650}=250'000 M^{-1}cm^{-1}), respectively and r is the ratio between absorbance at 280 nm and 650 nm of the pure dye (0.05 in the case of Cy5). This accounts for the non-neglibible absorbance of the pure dye at 280 nm.

SAF Immunoassay: Synthesis of NVOC-APTES

The synthesis of NVOC-APTES (4,5-Dimethoxy-2-nitro-benzyl[N-(3-triethoxysily1)-propyl]-carbamate) was performed according to a modified protocol described in Ref.[146][Fig. 7.5]. In brief, a solution of NVOC-Cl (Sigma-Aldrich, 3,4-dimethoxy-6-nitro-benzyl chloroformate, 0.10 mg; 0.36 mmol) and dry pyridine (1 ml) in dry toluene (5 ml) was heated to 40° in a dry 10 ml round bottom flask sealed with a rubber cap. The cap was punctured with the tip of a needle, which was left inside to depressurize. Upon addition of 3-aminopropyl triethoxysilane (0.13 g; 0.57 mmol) through a second needle a colorless precipitate immediately formed and the mixture was stirred at 40°C for 7 h. The precipitate was removed by decantation and the supernatant evaporated. Purification of the residue by chromatography, with 0.063-0.2 mm silicagel (Merck, Geneva, Switzerland) as stationary phase and n-hexane/AcOEt/Et$_3$N 7:3:0.1 as mobile phase, afforded the desired product as an orange/brown solid/oil (66.5 mg; 56.8%). The purity of the obtained NVOC-APTES was confirmed by ^{1}H-NMR spectroscopy. ^{1}H NMR (CDCl$_2$) δ : 7.70 (s, lH), 7.00 (s, lH), 5.50<s, 2H), 5.16 (br t, lH), 3.97 (s, 3H), 3.95 (s, 3H), 3.81 (q, J = 7.0 Hz, 6H), 3.22 (q, J = 6.6 Hz, 2H), 1.65 (m, 2H), 1.22 (t, J = 7.0 Hz, 9H), 0.64 (m, 2H).

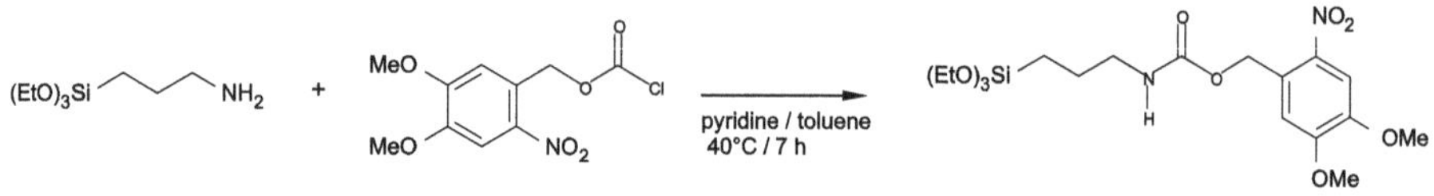

Figure 7.5: Reaction scheme for the synthesis of NVOC-APTES.

SAF Immunoassay: Surface chemistry with NVOC-APTES

For the coupling of NVOC-APTES to Zeonex® a difficulty was to find a solvent in which NVOC-APTES is soluble and to which the substrate is resistant. Dimethyl formamide (DMF) provided these properties. The coupling to Zeonex® was performed by prehydrolyzing NVOC-APTES (10 mg/ml in DMF) with 1 mM NaOH for 1 h. The solution was then reacted with oxygen-plasma treated Zeonex® (refer to 7) for 6 h. The polymer was rinsed extensively and sonicated in DMF to remove non-covalently bound NVOC-APTES.

8 Summary

In conclusion, simultanesous near- and far-field microscopy has opened up new possibilities in sub-diffraction imaging and the measurement of fast processes at membranes and surfaces. These applications were extended to the single-molecule level with the development of a second generation microscope system with single-molecule sensitivity for both SAF and UAF emission modes. The centerpiece of the microscope, the NA 1.0 2-*Theta* objective, provides a 2-fold improvement of the lateral image resolution, a 3-fold enhancement of the UAF collection efficiency compared to the prototype optics and can be used with a range of wavelengths.

With 3D-SAF microscopy a generic concept has been introduced for surface-selective, non-invasive 3D imaging with nanometer resolution in the historically more difficult dimension – along the optical axis. Several technically feasible approaches are proposed to overcome the current limitation of the diffraction-limited lateral resolution. These include STED or single-molecule localization of photoswitchable probes.

The SAF approach has been shown to be very powerful for surface-confined FCS in that it provides a small excitation spot of Gaussian shape, thereby minimimizing out-of-focus photobleaching and simplifying the interpretation of FCS curves. For the first time FCS has been shown using both SAF and UAF simultaneously. FCS measurements and simulations of free diffusion in solution have demonstrated how well defined the detection volumes for SAF and UAF are. Parallel FCS can provide the basis for measuring binding equilibria, rate constants, axial diffusion components, or axial concentration profiles at surfaces. The axial sensitivity provided by the parallel detection of SAF and UAF is the basis of a newly introduced FCS scheme for recognizing sample-related artefacts in the determination of diffusion coefficients in cell membranes. It was shown that FCS curves flawed by contributions of membrane geometry can be identified clearly better than in standard confocal FCS. The method can be combined with approaches to additionally minimize optical artefacts for even more accurate diffusion measurements.

The simplicity of the optical geometry for the collection of SAF has promoted the development of an inexpensive and easy-to-use immunoassay platform for the rapid and sensitive quantification of bioanalytes. Three commercially interesting assays were presented with picomolar sensitivity in

CHAPTER 8. SUMMARY

the turn of only thirteen minutes. The developed assay technology has great potential of replacing the work-intensive and time-consuming ELISA, providing a comparable sensitivity and dynamic range, requiring only a single step, being much faster, and using only a fraction of the amount of substances. The portable SAF immunoassay platform combines high detection performance with low cost and has the capability of bringing sensitive testing to the point-of-care.

Appendix

Theory of Fluorescence Emission at Dielectric Interfaces

The theory describing the angular emission properties of a dipole near a RI discontinuity was adopted from Ref.[43] where a fluorescing molecule is described in a completely classical electrodynamic framework as a radiating electric dipole. The calculation of the angular distribution of radiation is described in the following.

In the case of a dipole orientation orthogonal to the surface the flux intensity S into the positive semispace ($z>0$) into the solid angle element $d\Omega^2$ as a function of the distance z of the emitter in the medium with RI n_1 from the medium with RI n_2 is given by

$$\frac{d^2S}{d\Omega^2} = n_1 sin^2(\theta) \left|1 + R_p e^{iw_1 z}\right|^2, \tag{8.1}$$

and the flux intensity into the negative semispace ($z<0$) is given by

$$\frac{d^2S}{d\Omega^2} = \frac{n_2 w_2 q}{n_1^2 |w_1|} T_p^2 e^{-2\Im(w_1)z}. \tag{8.2}$$

In the case of a dipole orientation parallel to the interface the flux intensity S into the positive semispace one has[1]

$$\frac{d^2S}{d\Omega^2} = \frac{w_1^2}{n_1} \left\{ cos^2(\phi) \left|1 - R_p e^{2iw_1 z}\right|^2 + n_1 sin^2(\phi) \left|1 + R_s e^{2iw_1 z}\right|^2 \right\}, \tag{8.3}$$

and into the negative semispace one has

$$\frac{d^2S}{d\Omega^2} = \frac{n_2 w_2^2}{|w_1|^2} \left\{ cos^2(\phi) |T_s|^2 + \frac{|w_1|^2}{n_1^2} sin^2(\phi) |T_p|^2 \right\} e^{-2\Im(w_1)z}, \tag{8.4}$$

---

[1]It is to be noted that the exponential factor $e^{2iw_1 z}$ was accidentally left out in the original publication[43] and was added upon consultation of the main author Prof. Dr. Jörg Enderlein.

APPENDIX

using the reflection and the transmission coefficients for plane p and s waves given by

$$R_p = \frac{w_1 n_2^2 - w_2 n_1^2}{w_1 n_2^2 + w_2 n_1^2} \tag{8.5}$$

$$R_s = \frac{w_1 - w_2}{w_1 + w_2} \tag{8.6}$$

$$T_p = \frac{2 n_1 n_2 w_1}{w_1 n_2^2 + w_2 n_1^2} \tag{8.7}$$

$$T_s = \frac{2 w_1 w_2}{w_1 + w_2}, \tag{8.8}$$

with $w_1 = \sqrt{n_1^2 - q^2}$, $w_2 = \sqrt{n_2^2 - q^2}$ and $q = n_1 \sin(\theta)$ for $z > 0$, $q = n_2 \sin(\theta)$ for $z < 0$. For a randomized dipole orientation the radiation flux is calculated as the sum of the orthogonal and parallel contributions with the orthogonal weighted 1/3 and the parallel 2/3. This applies for fluorphores diffusing freely in solution or rotating around a flexible chemical linker where the rotational correlation times can be regarded as being much faster than the excited state lifetime.

Collection efficiency

From the angular distribution of radiation the CE of a conventional microscope objective in dependence of the NA for a molecule at z=0 is calculated as[43]

$$CE = 2\pi \int_{\theta_{min}}^{\pi} \sin(\theta) \left(\frac{d^2 S_*}{d\Omega^2} \right) d\theta, \tag{8.9}$$

using the lower integration limit

$$\theta_{min} = \pi - \arcsin(NA/n_g), \tag{8.10}$$

with n_g being the RI of the glass/immersion oil (here 1.52) and $d^2 S_*$ the flux intensity of a randomized dipole into the negative semispace (glass). For normalization the integral is divided by the total flux. This calculation supposes no optical losses occuring at lens interfaces.

References

[1] Verdes, D., Ruckstuhl, T., and Seeger, S. *J. Biomed. Opt.* **12**(3), 034012 (2007).

[2] Hecht, B., Bielefeldt, H., Pohl, D. W., Novotny, L., and Heinzelmann, H. *J. Appl. Phys.* **84**(11), 5873–5882 (1998).

[3] Pavani, S. R. P., Thompson, M. A., Biteen, J. S., Lord, S. J., Liu, N., Twieg, R. J., Piestun, R., and Moerner, W. E. *Proc. Natl. Acad. Sci. U.S.A.* **106**(9), 2995–2999 (2009).

[4] Schmidt, R., Wurm, C. A., Jakobs, S., Engelhardt, J., Egner, A., and Hell, S. W. *Nat. Methods* **5**(6), 539–544 (2008).

[5] Shtengel, G., Galbraith, J. A., Galbraith, C. G., Lippincott-Schwartz, J., Gillette, J. M., Manley, S., Sougrat, R., Waterman, C. M., Kanchanawong, P., Davidson, M. W., Fetter, R. D., and Hess, H. F. *Proc. Natl. Acad. Sci. U.S.A.* **106**(9), 3125–3130 (2009).

[6] Juette, M. F., Gould, T. J., Lessard, M. D., Mlodzianoski, M. J., Nagpure, B. S., Bennett, B. T., Hess, S. T., and Bewersdorf, J. *Nat. Methods* **5**(6), 527–529 (2008).

[7] Huang, B., Wang, W., Bates, M., and Zhuang, X. *Science* **319**(5864), 810–813 (2008).

[8] Oshikane, Y., Kataoka, T., Okuda, M., Hara, S., Inoue, H., and Nakano, M. *Sci. Technol. Adv. Mater.* **8**(3), 181–185 (2007).

[9] Sarkar, A., Robertson, R. B., and Fernandez, J. M. *Proc. Natl. Acad. Sci. U.S.A.* **101**(35), 12882–12886 (2004).

[10] Saffarian, S. and Kirchhausen, T. *Biophys. J.* **94**(6), 2333–2342 (2008).

[11] Samiee, K., Moran-Mirabal, J., Cheung, Y., and Craighead, H. *Biophys. J.* **90**(9), 3288–3299 (2006).

[12] Petra, S. *Biophys. J.* **85**(5), 2783–2784 (2003).

[13] Hassler, K., Leutenegger, M., Rigler, P., Rao, R., Rigler, R., Gösch, M., and Lasser, T. *Opt. Express* **13**(19), 7415–7423 (2005).

[14] Ries, J., Petrov, E. P., and Schwille, P. *Biophys. J.* **95**(1), 390–399 (2008).

[15] Ries, J., Ruckstuhl, T., Verdes, D., and Schwille, P. *Biophys. J.* **94**(1), 221 – 229 (2008).

[16] Ruckstuhl, T., Rankl, M., and Seeger, S. *Biosens. Bioelectron.* **18**(9), 1193 – 1199 (2003).

[17] Krieg, A., Laib, S., Ruckstuhl, T., and Seeger, S. *ChemBioChem* **4**(7), 589–592 (2003).

[18] Krieg, A., Laib, S., Ruckstuhl, T., and Seeger, S. *ChemBioChem* **5**(12), 1680–1685 (2004).

[19] Krieg, A., Ruckstuhl, T., and Seeger, S. *Anal. Biochem.* **349**(2), 181–5 (2006).

[20] Laib, S., Krieg, A., Rankl, M., and Seeger, S. *Appl. Surf. Sci.* **252**(22), 7788–7793 (2006).

[21] Rabe, M., Verdes, D., Rankl, M., Artus, G. R. J., and Seeger, S. *ChemPhysChem* **8**(6), 862–872 (2007).

[22] Rabe, M., Verdes, D., Zimmermann, J., and Seeger, S. *J. Phys. Chem. B* **112**(44), 13971–13980 (2008).

[23] Rabe, M., Verdes, D., and Seeger, S. *Soft Matter* **5**(5), 1039–1047 (2009).

[24] Kurzbuch, D., Bakker, J., Melin, J., Jönsson, C., Ruckstuhl, T., and MacCraith, B. *Sensor. Actuat. B-Chem.* **137**(1), 1 – 6 (2009).

[25] Rabe, M., Verdes, D., and Seeger, S. *J. Phys. Chem. B* **114**(17), 5862–5869 (2010).

[26] Rabe, M., Verdes, D., and Seeger, S. *Adv. Colloid Interface Sci.* **162**(1âĂŞ2), 87–106 February (2011).

[27] Tang, D., Yuan, R., and Chai, Y. *Analyst* **133**(7), 933–938 (2008).

[28] Tan, W., Huang, Y., Nan, T., Xue, C., Li, Z., Zhang, Q., and Wang, B. *Anal. Chem.* **82**(2), 615–620 (2009).

[29] Zheng, G., Patolsky, F., Cui, Y., Wang, W. U., and Lieber, C. M. *Nat. Biotech.* **23**(10), 1294–1301 (2005).

[30] Xu, J., Reiserer, R., Tellinghuisen, J., Wikswo, J. P., and Baudenbacher, F. J. *Anal. Chem.* **80**(8), 2728–2733 (2008).

[31] Kurita, R., Yokota, Y., Sato, Y., Mizutani, F., and Niwa, O. *Anal. Chem.* **78**(15), 5525–5531 (2006).

[32] Mauriz, E., Calle, A., Manclús, J., Montoya, A., and Lechuga, L. *Anal. Bioanal. Chem.* **387**, 1449–1458 (2007).

[33] Glick, S. *Nature* **474**(7353), 580–580 (2011).

[34] Kukar, T., Eckenrode, S., Gu, Y., Lian, W., Megginson, M., She, J.-X., and Wu, D. *Anal. Biochem.* **306**(1), 50–54 (2002).

[35] Kerman, K., Endo, T., Tsukamoto, M., Chikae, M., Takamura, Y., and Tamiya, E. *Talanta* **71**(4), 1494 – 1499 (2007).

[36] Rissin, D. M., Kan, C. W., Campbell, T. G., Howes, S. C., Fournier, D. R., Song, L., Piech, T., Patel, P. P., Chang, L., Rivnak, A. J., Ferrell, E. P., Randall, J. D., Provuncher, G. K., Walt, D. R., and Duffy, D. C. *Nat. Biotech.* **28**(6), 595–599 (2010).

[37] Ni, J., Lipert, R. J., Dawson, G. B., and Porter, M. D. *Anal. Chem.* **71**(21), 4903–4908 (1999).

[38] Deiss, F., LaFratta, C. N., Symer, M., Blicharz, T. M., Sojic, N., and Walt, D. R. *J. Am. Chem. Soc.* **131**(17), 6088–6089 (2009).

[39] Li, M., Sun, Y., Chen, L., Li, L., Zou, G., Zhang, X., and Jin, W. *Electroanal.* **22**(3), 333–337 (2010).

[40] http://www.microscopyu.com/articles/fluorescence/tirf/tirfintro.html, (1. April 2012).

[41] Hellen, E. H. and Axelrod, D. *J. Opt. Soc. Am. B* **4**(3), 337–350 (1987).

[42] Lukosz, W. *J. Opt. Soc. Am.* **69**(11), 1495–1503 (1979).

[43] Enderlein, J., Ruckstuhl, T., and Seeger, S. *Appl. Opt.* **38**(4), 724–732 (1999).

[44] Hecht, B., Sick, B., Wild, U. P., Deckert, V., Zenobi, R., Martin, O. J. F., and Pohl, D. W. *J. Chem. Phys.* **112**(18), 7761–7774 (2000).

[45] Zenobi, R. and Deckert, V. *Angew. Chem.* **112**(10), 1814–1825 (2000).

[46] Levene, M. J., Korlach, J., Turner, S. W., Foquet, M., Craighead, H. G., and Webb, W. W. *Science* **299**(5607), 682–686 (2003).

[47] Hell, S. W. *Science* **316**(5828), 1153–1158 (2007).

[48] Huang, B., Bates, M., and Zhuang, X. *Annu. Rev. Biochem.* **78**(1), 993–1016 (2009).

[49] Schermelleh, L., Heintzmann, R., and Leonhardt, H. *J. Cell. Biol.* **190**(2), 165–175 (2010).

[50] Hell, S. W. and Wichmann, J. *Opt. Lett.* **19**(11), 780–782 (1994).

[51] Hell, S. W. and Kroug, M. *Appl. Phys. B-Lasers O.* **60**(5), 495–497 (1995-05-01).

[52] Gustafsson, M. G. L. *J. Microsc.* **198**(2), 82–87 (2000).

[53] Heintzmann, R., Jovin, T. M., and Cremer, C. *J. Opt. Soc. Am. A* **19**(8), 1599–1609 (2002).

[54] Dertinger, T., Colyer, R., Iyer, G., Weiss, S., and Enderlein, J. *Proc. Natl. Acad. Sci. U.S.A.* **106**(52), 22287–92 (2009).

[55] Hess, S., Girirajan, T., and Mason, M. *Biophys. J.* **91**(11), 4258–4272 (2006).

[56] Rust, M. J., Bates, M., and Zhuang, X. *Nat. Methods* **3**(10), 793–796 (2006).

[57] Betzig, E., Patterson, G. H., Sougrat, R., Lindwasser, O. W., Olenych, S., Bonifacino, J. S., Davidson, M. W., Lippincott-Schwartz, J., and Hess, H. F. *Science* **313**(5793), 1642–1645 (2006).

[58] Folling, J., Bossi, M., Bock, H., Medda, R., Wurm, C. A., Hein, B., Jakobs, S., Eggeling, C., and Hell, S. W. *Nat. Methods* **5**(11), 943–945 (2008).

[59] Heilemann, M., van de Linde, S., Schüttpelz, M., Kasper, R., Seefeldt, B., Mukherjee, A., Tinnefeld, P., and Sauer, M. *Angew. Chem. Int. Edit.* **47**(33), 6172–6176 (2008).

[60] Müller, C. B. and Enderlein, J. *Phys. Rev. Lett.* **104**(19), 198101 (2010).

[61] Enderlein, J. *Appl. Phys. Lett.* **87**(9), 094105–3 (2005).

[62] Donnert, G., Keller, J., Wurm, C. A., Rizzoli, S. O., Westphal, V., Schönle, A., Jahn, R., Jakobs, S., Eggeling, C., and Hell, S. W. *Biophys. J.* **92**(8), 67–69 (2007).

[63] Kuhle, J., Regeniter, A., Leppert, D., Mehling, M., Kappos, L., Lindberg, R. L., and Petzold, A. *J. Neuroimmunol.* **220**(1–2), 114–119 (2010).

[64] Bonham, A. J., Neumann, T., Tirrell, M., and Reich, N. O. *Nucleic Acids Res.* **37**(13), 94–104 (2009).

[65] Bruls, D. M., Evers, T. H., Kahlman, J. A. H., van Lankvelt, P. J. W., Ovsyanko, M., Pelssers, E. G. M., Schleipen, J. J. H. B., de Theije, F. K., Verschuren, C. A., van der Wijk, T., van Zon, J. B. A., Dittmer, W. U., Immink, A. H. J., Nieuwenhuis, J. H., and Prins, M. W. J. *Lab Chip* **9**, 3504–3510 (2009).

[66] Teramura, Y., Arima, Y., and Iwata, H. *Anal. Biochem.* **357**(2), 208–215 (2006).

[67] Wu, G., Datar, R. H., Hansen, K. M., Thundat, T., Cote, R. J., and Majumdar, A. *Nat Biotech* **19**(9), 856–860 (2001).

[68] Kurosawa, S., Aizawa, H., and Park, J.-W. *Analyst* **130**(11), 1495–1501 (2005).

[69] Ruckstuhl, T. and Verdes, D. *Opt. Express* **12**(18), 4246–4254 (2004).

[70] Ruckstuhl, T., Verdes, D., Winterflood, C. M., and Seeger, S. *Opt. Express* **19**(7), 6836–6844 (2011).

[71] Ruckstuhl, T., Winterflood, C. M., and Seeger, S. *Anal. Chem.* **83**(6), 2345–2350 (2011).

[72] Winterflood, C. M., Ruckstuhl, T., Verdes, D., and Seeger, S. *Phys. Rev. Lett.* **105**(10), 108103 (2010).

[73] Winterflood, C. M., Ruckstuhl, T., Reynolds, N. P., and Seeger, S. *ChemPhysChem* **13**(16), 3655–3660 (2012).

[74] Ruckstuhl, T., Enderlein, J., Jung, S., and Seeger, S. *Anal. Chem.* **72**(9), 2117–2123 (2000).

[75] Novotny, L. *J. Opt. Soc. Am. A* **14**(1), 91–104 (1997).

[76] Mattheyses, A. L. and Axelrod, D. *J. Biomed. Opt.* **11**(1) (2006).

[77] Seiffert, S. and Oppermann, W. *Polymer* **49**(19), 4115–4126 (2008).

[78] Lakowicz, J. *Principles of Fluorescence Spectroscopy, 3rd Ed.* Kluwer Academic/Plenum Publishers, New York, Boston, Dordrecht, London, Moscow, (2006).

[79] Wahl, M., Gregor, I., Patting, M., and Enderlein, J. *Opt. Express* **11**(26), 3583–3591 (2003).

[80] Hess, S. T. and Webb, W. W. *Biophys. J.* **83**(4), 2300–2317 (2002).

[81] Leutenegger, M. *Single Molecule Detection on Surfaces*. PhD thesis, École Polytechnique Fédérale de Lausanne, Lausanne, Germany, (2007).

[82] Widengren, J., Mets, U., and Rigler, R. *J. Phys. Chem.* **99**(36), 13368–13379 (1995).

[83] Gregor, I., Patra, D., and Enderlein, J. *ChemPhysChem* **6**(1), 164–70– (2005).

[84] Koppel, D. E. *Phys. Rev. A* **10**, 1938–1945 (1974).

[85] Zhao, M., Jin, L., Chen, B., Ding, Y., Ma, H., and Chen, D. *Appl. Opt.* **42**(19), 4031–4036 (2003).

[86] Rüttinger, S. and Koberling, F. *Picoquant Technical Note* (2011).

[87] Ries, J. *Advanced Fluorescence Correlation Techniques to Study Membrane Dynamics.* PhD thesis, TU Dresden, Dresden, Germany, (2008).

[88] Richards, B. and Wolf, E. *Proc. R. Soc. London Ser. A* **253**(1274), 358–379 (1959).

[89] Ling, H. and Lee, S.-W. *J. Opt. Soc. Am. A* **1**(9), 965–973 (1984).

[90] Török, P., Varga, P., Laczik, Z., and Booker, G. R. *J. Opt. Soc. Am. A* **12**(2), 325–332 (1995).

[91] Novotny, L. *Lecture Notes on Nano-optics.* (Rochester Univ., NY, 2000).

[92] Marrocco, M. *Chem. Phys. Lett.* **449**(1-3), 227 – 230 (2007).

[93] Buschmann, V., Krämer, B., Koberling, F., Macdonald, R., and Rüttinger, S. *Picoquant Application Note* (2009).

[94] Dertinger, T., Pacheco, V., von der Hocht, I., Hartmann, R., Gregor, I., and Enderlein, J. *ChemPhysChem* **8**(3), 433–443 (2007).

[95] Blom, H., Hassler, K., Chmyrov, A., and Widengren, J. *Int. J. Mol. Sci.* **11**(2), 386–406 (2010).

[96] Axelrod, D., Koppel, D., Schlessinger, J., Elson, E., and Webb, W. *Biophys. J.* **16**(9), 1055–1069 (1976).

[97] Barak, L. and Webb, W. *J. Cell Biol.* **95**(3), 846–852 (1982).

[98] Meyer, T. and Schindler, H. *Biophys. J.* **54**(6), 983–93– (1988).

[99] Reynolds, N. P., Soragni, A., Rabe, M., Verdes, D., Liverani, E., Handschin, S., Riek, R., and Seeger, S. *J. Am. Chem. Soc.* **133**(48), 19366–19375 (2011).

[100] Yguerabide, J. and Stryer, L. *Proc. Natl. Acad. Sci. U.S.A.* **68**(6), 1217–1221 (1971).

[101] Axelrod, D. *Biophys. J.* **26**(3), 557–573 (1979).

[102] Badley, R. A., Martin, W. G., and Schneider, H. *Biochemistry* **12**(2), 268–275 (1973).

[103] Aizenbud, B. and Gershon, N. *Biophys. J.* **48**(4), 543–546 (1985).

[104] Hall, D. *Anal. Biochem.* **377**(1), 24–32 (2008).

[105] Weiss, M., Hashimoto, H., and Nilsson, T. *Biophys. J.* **84**(6), 4043–4052 (2003).

[106] Milon, S., Hovius, R., Vogel, H., and Wohland, T. *Chemical Physics* **288**(2–3), 171–186 (2003).

[107] Ries, J. and Schwille, P. *Phys. Chem. Chem. Phys.* **10**(24), 3487–3497 (2008).

[108] Rotsch, C., Jacobson, K., and Radmacher, M. *Proc. Natl. Acad. Sci. U.S.A.* **96**(3), 921–926 (1999).

[109] de Laat, S., van der Saag, P., Elson, E., and Schlessinger, J. *Proc. Natl. Acad. Sci. U.S.A.* **77**(3), 1526–8– (1980).

[110] Thompson, T. E. and Tillack, T. W. *Annu. Rev. Biophys. Bio.* **14**, 361–386 (1985).

[111] Wieser, S., Weghuber, J., Sams, M., Stockinger, H., and Schutz, G. J. *Soft Matter* **5**(17), 3287–3294 (2009).

[112] Przybylo, M., Sykora, J., Humpolickova, J., Benda, A., Zan, A., and Hof, M. *Langmuir* **22**(22), 9096–9 (2006).

[113] Hetzer, M., Heinz, S., Grage, S., and Bayerl, T. M. *Langmuir* **14**(5), 982–984 (1998).

[114] Zhang, L. and Granick, S. *J. Chem. Phys.* **123**(21), 211104–4 (2005).

[115] Stottrup, B. L., Veatch, S. L., and Keller, S. L. *Biophys. J.* **86**(5), 2942–2950 (2004).

[116] Dertinger, T., von der Hocht, I., Benda, A., Hof, M., and Enderlein, J. *Langmuir* **22**(22), 9339–9344 (2006).

[117] Axelrod, D. *J. Biomed. Opt.* **6**(1), 6–13 (2001).

[118] Barroca, T., Balaa, K., Delahaye, J., Lévêque-Fort, S., and Fort, E. *Opt. Lett.* **36**(16), 3051–3053 (2011).

[119] Barroca, T. S., Fort, E., Balaa, K., Lévêque-Fort, S., and Fort, E. *Phys. Rev. Lett* (2012).

[120] Enderlein, J., Toprak, E., and Selvin, P. R. *Opt. Express* **14**(18), 8111–8120 (2006).

[121] Böhmer, M. and Enderlein, J. *J. Opt. Soc. Am. B* **20**(3), 554–559 (2003).

[122] Briddon, S. J., Middleton, R. J., Cordeaux, Y., Flavin, F. M., Weinstein, J. A., George, M. W., Kellam, B., Hill, S. J., and Black, J. W. *Proc. Natl. Acad. Sci. U.S.A.* **101**(13), 4673–4678 (2004).

[123] Cordeaux, Y., Briddon, S. J., Alexander, S. P. H., Kellam, B., and Hill, S. J. *FASEB J.* **22**(3), 850–860 (2008).

[124] Gerken, M., Krippner-Heidenreich, A., Steinert, S., Willi, S., Neugart, F., Zappe, A., Wrachtrup, J., Tietz, C., and Scheurich, P. *BBA Biomembranes* **1798**(6), 1081–1089 (2010).

[125] Enderlein, J., Gregor, I., Patra, D., and Fitter, J. *Curr. Pharm. Biotechnol.* **5**(2), 155–161 (2004).

[126] Ruan, Q., Cheng, M. A., Levi, M., Gratton, E., and Mantulin, W. W. *Biophys. J.* **87**(2), 1260 – 1267 (2004).

[127] Ries, J. and Schwille, P. *Biophys. J.* **91**(5), 1915 – 1924 (2006).

[128] Ries, J., Chiantia, S., and Schwille, P. *Biophys. J.* **96**(5), 1999 – 2008 (2009).

[129] Wawrezinieck, L., Rigneault, H., Marguet, D., and Lenne, P.-F. *Biophys. J.* **89**(6), 4029 – 4042 (2005).

[130] Kastrup, L., Blom, H., Eggeling, C., and Hell, S. W. *Phys. Rev. Lett.* **94**, 178104 (2005).

[131] Eggeling, C., Ringemann, C., Medda, R., Schwarzmann, G., Sandhoff, K., Polyakova, S., Belov, V. N., Hein, B., von Middendorff, C., Schonle, A., and Hell, S. W. *Nature* **457**(7233), 1159–1162 (2009).

[132] Winterflood, C. M., Ruckstuhl, T., and Seeger, S. *Biosensors* **3**(1), 108–115 (2013).

[133] Jonsson, C., Aronsson, M., Rundstrom, G., Pettersson, C., Mendel-Hartvig, I., Bakker, J., Martinsson, E., Liedberg, B., MacCraith, B., Ohman, O., and Melin, J. *Lab Chip* **8**, 1191–1197 (2008).

[134] Diaz-Quijada, G. A., Peytavi, R., Nantel, A., Roy, E., Bergeron, M. G., Dumoulin, M. M., and Veres, T. *Lab Chip* **7**, 856–862 (2007).

[135] Pai, M., Riley, L. W., and Jr, J. M. C. *Lancet Infect. Dis.* **4**(12), 761 – 776 (2004).

[136] Tamborrini, G., Krebs, A., Michel, M., Michel, B., and Ciurea, A. *Zeitschr. f. Rheumat.* **70**, 154–159 (2011).

[137] Carter, A. B. and Howanitz, P. J. *Arch. Pathol. Lab. Med.* **127**(11), 1424–1442 (2003).

[138] Fodor, S., Read, J., Pirrung, M., Stryer, L., Lu, A., and Solas, D. *Science* **251**(4995), 767–773 (1991).

[139] Jonas, U., del Campo, A., Kruger, C., Glasser, G., and Boos, D. *Proc. Natl. Acad. Sci. U.S.A.* **99**(8), 5034–9– (2002).

[140] Boos, D. *Synthesis and Characterisation of Photoreactive Silanes for the Self-Assembly on Functionalised Silica Surfaces*. PhD thesis, University of Mainz, Mainz, Germany, (2004).

[141] Stegmaier, P. *Surfaces for Functional and Patterned Immobilization of Proteins*. PhD thesis, Max Planck Institute, Stuttgart, Germany, (2007).

[142] Ruckstuhl, T. and Seeger, S., International Patent Application WO2008132247 (2008).

[143] Semenova, I. and Rodionov, V. In *Methods in Molecular Medicine*, volume 137, 93–102. (2007).

[144] Yamazaki, M. *J. Mol. Catal. A: Chem.* **213**(1), 81–87 (2004).

[145] Southwick, P. L., Ernst, L. A., Tauriello, E. W., Parker, S. R., Mujumdar, R. B., Mujumdar, S. R., Clever, H. A., and Waggoner, A. S. *Cytometry* **11**(3), 418–430 (1990).

[146] Jennane, J., Boutros, T., and Giasson, R. *Can. J. Chem.* **74**(12), 2509–2517 (1996).

List of Abbreviations

$3D-SAFM$ Three-dimensional supercritical angle fluorescence microscopy

aa amino acid

ACF Autocorrelation function

$APTES$ 3-aminopropyl triethoxysilane

CE Collection efficiency

cpm Count rate per molecule

CV Coefficient of variation

$DiIC_{18}(5)$ 1,1'-dioctadecyl-3,3,3',3'-tetramethylindodicarbocyanine

$DMEM$ Dulbecco's modified eagle medium

DMF Dimethyl formamide

$DOPC$ 1,2-dioleoyl-sn-glycero-3-phosphocholine

$DOPS$ 1,2-dioleoyl-sn-glycero-3-phospho-L-serine

ECL Electrochemiluminescence

$EDTA$ Ethylenediaminetetraacetic acid

$EGTA$ Ethylene glycol tetraacetic acid

$ELISA$ Enzyme-linked immunosorbent assay

$f-TIR$ Frustrated total internal reflection

FCS Fluorescence correlation spectroscopy

FRAP Fluorescence recovery after photobleaching

FWHM Full width at half maximum

HEPES 2-(4-(2-Hydroxyethyl)- 1-piperazinyl)-ethansulfonic acid

$IFN-\gamma$ Interferon-γ

$IL-2$ Interleukin-2

iPTH Intact parathyroid hormone

LOD Limit of detection

mB Molecular brightness

NA Numerical aperture

NHS N-hydroxy succinimidyl ester

NSOM Near-field scanning optical microscopy

NVOC Nitroveratryloxycarbonyl

PBS Phosphate buffered saline

PIPES Piperazine-N,N'-bis(2-ethanesulfonic acid)

PSF Point-spread function

PTH Parathyroid hormone

$SA-FCS$ Supercritical angle fluorescence correlation spectroscopy

SAF Supercritical angle fluorescence

SLB Supported lipid bilayers

SPAD Single-photon avalanche diode

SPR Surface plasmon resonance

SPT Single particle tracking

$TIR-FCS$ Total internal reflection fluorescence correlation spectroscopy

TPA Tripropylamine

UA − FCS Undercritical angle fluorescence correlation spectroscopy

UAF Undercritical angle fluorescence

ZMW Zero-mode wave guide

List of Figures

2.1	Principle of TIRF microscopy	6
2.2	Angular emission distribution and dipole orientation at an interface	7
2.3	Distance dependence of the angular emission distribution at an interface	8
2.4	Comparison between the surface selectivity of SAF and TIRF	9
2.5	Principle of transmission mode NSOM	9
2.6	Zero mode waveguides	10
2.7	Far-field techniques: Wide-field and confocal microscopy	11
2.8	Principle of STED microscopy	12
2.9	Principle of single-molecule localization microscopy	13
3.1	Principle of electrochemiluminescence	16
3.2	TIRF array biosensor	16
3.3	f-TIR array biosensor	17
3.4	Nanowire field-effect transistor	18
3.5	Principle of surface plasmon resonance	18
3.6	Microcantilever immunosensor	19
3.7	Quartz crystal microbalance immunosensor	19
4.1	Prototype parallel near- and far-field fluorescence microscope system	22
4.2	Collection efficiency and numerical aperture.	23
4.3	The *2-Theta* microscope objective.	24
4.4	Wide-field imaging with the *2-Theta* microscope objective.	25
4.5	Dependence of the SAF collection efficiency on the angular cut-off.	26
4.6	Schmematic and photos of the second generation *2-Theta* microscope	29
4.7	Lateral positioning of the excitation beam.	30
4.8	Intensity images of SAF and UAF in their image planes	30
4.9	Alignment of the angular aperture for the *2-Theta* objective	31

4.10 Lateral PSF of the *2-Theta* objective 32
4.11 Single-molecule imaging with the *2-Theta* objective 32
4.12 Output of the parabolic collector 33
4.13 Schematic and photographs of the test tube and the optical substrate 34
4.14 SAF-tube: Emission at the solute/polymer interface 35
4.15 Schematics and photographs of the fluorescence tube-reader 36
4.16 Collection efficiency function of the SAF-tube 37

5.1 Schematic setup for parallel SAF and UAF microscopy 40
5.2 Distance dependence of I_{SAF}/I_{UAF} for the prototype microscope 41
5.3 Axial localization accuracy of 3D-SAFM 42
5.4 Measurement of the axial profile of a microsphere with 3D-SAFM 44
5.5 Nanometer precision axial localization of nanobeads by 3D-SAFM 46
5.6 UAF and SAF images of fibroblast cells 47
5.7 3D-SAFM imaging of the microtubule network of a cell 48
5.8 Axial localization of single molecules by 3D-SAFM. 49
5.9 Single-molecule bleaching and I_{SAF}/I_{UAF} 50
5.10 Orientation dependence of the I_{SAF}/I_{UAF} 51
5.11 Tracking of the axial diffusion of nanobeads 52
5.12 SAF and UAF intensity tracks of diffusing nanobeads 53
5.13 Autocorrelation functions for slow and fast diffusion. 55
5.14 Effect of triplet dynamics on the ACF. 57
5.15 Saturation curves of Atto655. 58
5.16 Afterpulsing correction of the ACF. 59
5.17 Comparison between the SAF and UAF ACFs for free diffusion 63
5.18 Comparison between simulated and experimental SAF- and UAF-CS of free diffusion 64
5.19 Comparison between simulated and experimental crosscorrelation functions for free
    diffusion 65
5.20 Effect of the SAF cut-off angle on the ACF of free diffusion 66
5.21 Imaging of the membrane diffusion of Cellmask^{TM} 67
5.22 SAF- and UAF-CS on a SLB 68
5.23 Orientation of Cellmask^{TM} in a SLB 69
5.24 Principle of parallel SAF- and UAF-CS of membrane diffusion 70
5.25 Parallel SAF- and UAF-CS of diffusion in a SLB 71
5.26 Imaging of the fibroblast cell membrane 73

5.27	FCS of membrane diffusion in fibroblast cells	74
5.28	Diffusion coefficients and Pearson's correlation coefficients	75
5.29	Imaging of the HeLa cell membrane	76
5.30	FCS of membrane diffusion in HeLa cells	76
6.1	Principle of the SAF-based immunoassay	82
6.2	Protein immobilization chemistry on Zeonex®	82
6.3	SAF-tube polymer autofluorescence reduction	83
6.4	Coating procedure and autoflourescence	84
6.5	IFN-gamma high-concentration range sandwich-assay	86
6.6	IFN-gamma low-concentration range sandwich-assay	87
6.7	IL-2 high-concentration range sandwich-assay	88
6.8	IL-2 low-concentration range sandwich-assay	88
6.9	PTH sandwich-assay	89
6.10	Effects of mass-transport on SAF-tube assays	90
6.11	Effects of capture-spot reduction on SAF-tube assays	92
6.12	Principle of photolithography with NVOC-APTES	93
6.13	NVOC-APTES UV spectra	93
6.14	Fluorescence image of photo-patterned Cy5	94
6.15	Parallelization of SAF-tube assays	95
7.1	Adsorption of Atto655 to the coverslip	99
7.2	Limit of detection	101
7.3	Chemical structure of Zeonex®	102
7.4	Antibody labelling reaction	102
7.5	Synthesis of NVOC-APTES	104

Acknowledgements

First and above all, I would like to express my deepest gratitude to Thomas Ruckstuhl, who supported me and taught me throughout my thesis with his brilliant ideas, exceptional expertise and relentless encouragment. He was a great support during my good and bad times, may it be personal or work-related. He is not merely a colleague, but has become a dear friend. I truly cherish the oppurtunity to have worked with him. Without him this thesis would never have been completed or written.

I would like to thank Dorinel Verdes for his continuous help and for introducing me to a field which was initially new to me. I could always turn to him in confindence for advice and I very much enjoyed his company.

I thank Ana Stojanovic and Nicholas Reynolds for their help when problems exceeded my knowledge and for the fun moments we spent inside and outside the institute.

Much appreciation goes to Roland Zehnder and Armin Kühne for their indespensable support in the area of machinery, tools and custom-made components.

Many thanks go to Chantal Henningsen-Conus for taking the load off of me when it came to administrative matters, for keeping me always well-informed and for running the show so smoothly. I will always remember her friendly smile and hello every day we met.

My thanks also go to all the group members for the pleasent time, namely Jan Zimmermann, Michael Rabe, Georg Artus, Qiang Li, Jungpin Zhang, Sandro Olveira, Georg Meseck, Simon Forster, and Isabelle Halstrick.

Finally, I am grateful to Stefan Seeger for having provided me the chance to work in his group on these absorbing projects.

Curriculum Vitae

Personal
Surname: Winterflood
Names: Christian Matthew
Date of birth: 24th March 1982
Citizenship: British/Swiss

Education and Research Experience

Sept. '12 - present: *Postdoctoral position*
Swiss Federal Institute of Technology (ETH), Zürich, Switzerland

May '08 - Sept. '12: *PhD Program in Physical Chemistry*
University of Zürich, Switzerland
PhD-Thesis: *"Supercritical angle fluorescence-based microscopy, correlation spectroscopy, and biosensing"*
Supervisor: Prof. Dr. Stefan Seeger

May '06 - Apr. '08: *MSc Program in Biochemistry*
University of Zürich, Switzerland
Masters-Thesis: *"Designed Ankyrin repeat proteins: Towards protein folding and a new spectroscopic ruler"*
Supervisor: Prof. Dr. Benjamin Schuler

Oct. '05 - Jul. '06: *BSc Program in Biochemistry*
University of Zürich, Switzerland
Bachelors-Thesis: *"Effects of salt bridges on the stability of native EB1 and their influence on its folding kinetics by differential scanning calorimetry and circular dichroism spectroscopy"*
Supervisor: PD Dr. Ilian Jelesarov

Oct. '03 - Jul. '05: *Basic Studies in Biology*
University of Zürich, Switzerland
Aug. '97 - Jan. '02: *High School*
Kantonsschule Riesbach, Modern Language Grammar School, Zürich, Switzerland

i want morebooks!

Buy your books fast and straightforward online - at one of world's fastest growing online book stores! Environmentally sound due to Print-on-Demand technologies.

Buy your books online at
www.get-morebooks.com

Kaufen Sie Ihre Bücher schnell und unkompliziert online – auf einer der am schnellsten wachsenden Buchhandelsplattformen weltweit! Dank Print-On-Demand umwelt- und ressourcenschonend produziert.

Bücher schneller online kaufen
www.morebooks.de

 VDM Verlagsservicegesellschaft mbH
Heinrich-Böcking-Str. 6-8　　Telefon: +49 681 3720 174　　info@vdm-vsg.de
D - 66121 Saarbrücken　　　Telefax: +49 681 3720 1749　　www.vdm-vsg.de

Printed by Books on Demand GmbH, Norderstedt / Germany